NUCLEAR POWER

TECHNICAL AND INSTITUTIONAL OPTIONS FOR THE FUTURE

Committee on Future Nuclear Power Development

Energy Engineering Board

Commission on Engineering and Technical Systems

National Research Council

NATIONAL ACADEMY PRESS

Washington, D.C. 1992

NOTICE: The project that is the subject of this report was approved by the Governing Board of the National Research Council, whose members are drawn from the councils of the National Academy of Sciences, the National Academy of Engineering, and the Institute of Medicine. The members of the committee responsible for the report were chosen for their special competencies and with regard for appropriate balance.

This report has been reviewed by a group other than the authors according to procedures approved by a Report Review Committee consisting of members of the National Academy of Sciences, the National Academy of Engineering, and the Institute of Medicine.

This is a report of work supported by Contract Number DE-FGO2-88NE37983/R from the U.S. Department of Energy to the National Academy of Sciences - National Research Council.

Library of Congress Cataloging-in-Publication Data

Nuclear Power : technical and institutional options for the future / Committee on Future Nuclear Power Development, Energy Engineering Board, Commission on Engineering and Technical Systems, National Research Council.
p. cm.
Includes bibliographical references.
ISBN 0-309-04395-6
1. Nuclear power. I. National Research Council (U.S.). Committee on Future Nuclear Power Development. II. National Research Council (U.S.). Commission on Engineering and Technical Systems.
TK1078.N86 1992
333.792'4--dc20 92-18493
CIP

Copies available from:
National Academy Press
2101 Constitution Avenue, N.W.
Washington, D.C. 20418

S634

Printed in the United States of America

First Printing, June 1992
Second Printing, August 1992

COMMITTEE ON FUTURE NUCLEAR POWER DEVELOPMENT

Chairman

JOHN F. AHEARNE, Executive Director, Sigma Xi, The Scientific Research Society, Research Triangle Park, North Carolina

Members

SAID I. ABDEL-KHALIK, Associate Director, Nuclear Engineering and Health Physics, Georgia Institute of Technology, Atlanta
ADOLF BIRKHOFER, Lehrstuhl für Reaktordynamik und Reaktorsicherheit, Technische Universität München, Munich, Germany
SOL BURSTEIN, former Vice Chairman, Wisconsin Electric Power Company, Milwaukee
RALPH C. CAVANAGH (Liaison to Energy Engineering Board), Senior Staff Attorney, Natural Resources Defense Council, San Francisco
BENJAMIN HUBERMAN, Consultant, Washington, D.C.
CHARLES R. IMBRECHT, Chairman, California Energy Commission, Sacramento
PETER T. JOHNSON, former Administrator, Bonneville Power Administration, Portland, Oregon
E. MARCIA KATZ, Associate Professor, Department of Nuclear Engineering, University of Tennessee, Knoxville
EDWARD H. KLEVANS, Head, Nuclear Engineering Department, Pennsylvania State University, University Park
JOHN W. LANDIS, Senior Vice President and Director, Stone & Webster Engineering Corporation, Boston, Massachusetts
TERRY R. LASH, former Director, Illinois Department of Nuclear Safety, Springfield
EUGENE W. MEYER, former Managing Director, Kidder, Peabody & Company, Inc., New York, New York
KENNETH J. NEMETH, Executive Director, Southern States Energy Board, Norcross, Georgia
DAVID OKRENT, Professor of Engineering and Applied Science, University of California at Los Angeles
ZACK T. PATE, President, Institute of Nuclear Power Operations, Atlanta, Georgia
HOWARD K. SHAPAR, Counsel, Shaw, Pittman, Potts & Trowbridge, Washington, D.C.
NEIL E. TODREAS, Professor, Department of Nuclear Engineering, Massachusetts Institute of Technology, Cambridge
GEORGES A. VENDRYES, former Director, Commissariat a l'Energie Atomique, France

RICHARD WILSON (Liaison to Energy Engineering Board), Mallinckrodt Professor of Physics, Harvard University, Cambridge, Massachusetts

National Research Council Staff

ARCHIE L. WOOD, Executive Director, Commission on Engineering and Technical Systems
MAHADEVAN (DEV) MANI, Director, Energy Engineering Board
NORMAN M. HALLER, Consultant and Study Director
THERESA M. FISHER, Study Administrative Assistant

ENERGY ENGINEERING BOARD

Chairman

JOHN A. TILLINGHAST, President, Tiltec, Portsmouth, New Hampshire

Members

DONALD B. ANTHONY, Vice President and Manager for Technology, Bechtel Corporation, Houston, Texas
RICHARD E. BALZHISER, President and Chief Executive Officer, Electric Power Research Institute, Palo Alto, California
BARBARA R. BARKOVICH, Partner, Barkovich and Yap, Consultants, Berkeley, California
JOHN A. CASAZZA, President, CSA Energy Consultants, Arlington, Virginia
RALPH C. CAVANAGH, Senior Staff Attorney, Natural Resources Defense Council, San Francisco, California
DAVID E. COLE, Director, Center for the Study of Automotive Transportation, University of Michigan, Ann Arbor
H. M. (HUB) HUBBARD, Resources for the Future, Washington, D.C.
CHARLES R. IMBRECHT, Chairman, California Energy Commission, Sacramento
CHARLES D. KOLSTAD, Associate Professor, Institute for Environmental Studies and Department of Economics, University of Illinois, Urbana
HENRY R. LINDEN, Max McGraw Professor of Energy & Power Engineering & Management, and Director, Energy and Power Center, Illinois Institute of Technology, Chicago
S. L. (CY) MEISEL, former Vice President, Research (retired), Mobil R&D Corporation, Princeton, New Jersey
DAVID L. MORRISON, Technical Director, Energy, Resource & Environmental Systems Division, The MITRE Corporation, McLean, Virginia

MARC H. ROSS, Professor, Physics Department, University of Michigan, Ann Arbor
MAXINE L. SAVITZ, Managing Director, Garrett Ceramic Component Division, Torrance, California
HAROLD H. SCHOBERT, Chairman, Fuel Sciences Program, Department of Materials Science and Engineering, The Pennsylvania State University, University Park
GLENN A. SCHURMAN, Vice President, Production (retired), Chevron Corporation, San Francisco, California
JON M. VEIGEL, President, Oak Ridge Associate Universities, Oak Ridge, Tennessee
BERTRAM WOLFE, Vice President and General Manager, GE Nuclear Energy, San Jose, California

Ex-officio Board Member

RICHARD WILSON, Mallinckrodt Professor of Physics, Harvard University, Cambridge, Massachusetts

Energy Engineering Board Staff

MAHADEVAN (DEV) MANI, Director
CLARENDON, SUSANNA, Administrative Assistant
KAMAL ARAJ, Senior Program Officer
JAMES J. ZUCCHETTO, Senior Program Officer
NORMAN M. HALLER, Consultant

Commission on Engineering and Technical Systems

ARCHIE L. WOOD, Executive Director
MARLENE BEAUDIN, Associate Executive Director

Technical Editor

CAROLETTA LOWE, Columbia, Maryland

The National Academy of Sciences is a private, nonprofit, self-perpetuating society of distinguished scholars engaged in scientific and engineering research, dedicated to the furtherance of science and technology and to their use for the general welfare. Upon the authority of the charter granted to it by the Congress in 1863, the Academy has a mandate that requires it to advise the federal government on scientific and technical matters. Dr. Frank Press is president of the National Academy of Sciences.

The National Academy of Engineering was established in 1964, under the charter of the National Academy of Sciences, as a parallel organization of outstanding engineers. It is autonomous in its administration and in the selection of its members, sharing with the National Academy of Sciences the responsibility for advising the federal government. The National Academy of Engineering also sponsors engineering programs aimed at meeting national needs, encourages education and research, and recognizes the superior achievements of engineers. Dr. Robert M. White is president of the National Academy of Engineering.

The Institute of Medicine was established in 1970 by the National Academy of Sciences to secure the services of eminent members of appropriate professions in the examination of policy matters pertaining to the health of the public. The Institute acts under the responsibility given to the National Academy of Sciences by its congressional charter to be an advisor to the federal government and, upon its own initiative, to identify issues of medical care, research, and education. Dr. Kenneth Shine is the president of the Institute of Medicine.

The National Research Council was organized by the National Academy of Sciences in 1916 to associate the broad community of science and technology with the Academy's purposes of furthering knowledge and advising the federal government. Functioning in accordance with general policies determined by the Academy, the council has become the principal operating agency of both the National Academy of Sciences and the National Academy of Engineering in providing services to the government, the public, and the scientific and engineering communities. The Council is administered jointly by both Academies and the Institute of Medicine. Dr. Frank Press and Dr. Robert M. White are chairman and vice chairman, respectively, of the National Research Council.

Preface

In the Conference Report (100-724) accompanying the fiscal 1989 appropriations bill for Energy and Water Development (H.R. 4567, Public Law 100-371), the U.S. Congress requested that the National Academy of Sciences conduct " . . . a critical comparative analysis . . . of the practical technological and institutional options for future nuclear power development and for the formulation of coherent policy alternatives to guide the Nation's nuclear power development." The Senate Appropriations Committee Report 100-381 entitled Energy and Water Development Appropriation Bill, 1989, which also accompanied the bill, noted:

> The [Senate Committee on Appropriations] believes that nuclear fission remains an important option for meeting our electric energy requirements and maintaining a balanced national energy policy. The Committee continues to strongly support the need for a responsive nuclear fission program, but finds the current civilian nuclear power reactor program to be a deficient aggregate of numerous reactor types and conceptual variations being developed without the guidance of well-defined strategic objectives. The Committee finds further that the future development and institutionalization of nuclear power development should be rethought, newly defined, and directed to be responsive to current and projected conditions.

In response to the congressional request, the National Academy of Sciences formed the Committee on Future Nuclear Power Development under the Energy Engineering Board of the National Research Council. The Committee's formal Statement of Task appears below:

> The committee will conduct a critical comparative analysis of the practical technological and institutional options for future nuclear power development and formulate coherent policy alternatives to guide the nation's nuclear power development. The Congressional intent in directing this study was that the future development and institutionalization of nuclear power development be rethought, newly defined, and directed to be responsive to current and projected conditions.
>
> In conducting this critical comparative analysis, the committee will undertake the following tasks:
>
> 1. The committee will identify the full range of practical technological options for the next generation of civilian nuclear power reactors.
> 2. The committee will develop criteria to evaluate these options. These criteria should reflect the extent to which the technologies are

likely to lend themselves to nuclear power plants that will exhibit characteristics such as the following:

- Safety in operation;
- Economy of operation;
- Suitability for the operational, institutional, financial, regulatory, and policy environments that are likely to prevail at the time such plants might be constructed;
- Amenability to efficient and predictable licensing;
- Environmental acceptability, both in day-to-day operations and with respect to the fuel cycles they employ; and
- Resistance to diversion of sensitive nuclear materials.

3. The committee will evaluate the technological options in terms of these criteria.
4. The committee will review and assess development approaches for the next generation of reactors, taking into account likely federal funding limitations. Particular emphasis will be put on approaches to establishing the level of safety that can be achieved by these reactors, to defining regulatory requirements likely to be imposed on these reactors, and to establishing, during the research, development, and demonstration program, the extent to which such regulatory requirements have been met. The committee will also consider the appropriate role for and level of private sector involvement in the development program.
5. The committee will assess, in light of the technological options and development approaches under consideration, the relevance of existing facilities of the Department of Energy that support civilian power reactor development. The committee will also consider the need for any new facilities.
6. The committee will address other aspects of the future civilian power development program necessary for completion of the broad purposes of this study.
7. Based on the results of the foregoing tasks, the committee will develop a set of coherent policy alternatives to guide the nation's future civilian nuclear power development program. The committee will formulate recommendations in terms of these alternatives.
8. The committee will document its conclusions and recommendations, and the reasoning therefore, in a final report.

The Committee was constituted to reflect a wide range of expertise and views in order to ensure that the study's conclusions and recommendations would take into account all relevant considerations and demonstrate the greatest balance possible.

The full Committee met on eight occasions from May 1989 through March 1991. Subcommittees and working groups also convened during this period to review the literature and other relevant materials provided by government, industry, academic, and public interest organizations, and to prepare working drafts. During the study, the Committee was briefed by representatives from many organizations, including the Executive Branch (the Department of Energy and the national laboratories); the Congress (the Senate Committee on Energy and Natural Resources and the Office of Technology Assessment); the Nuclear Regulatory Commission; domestic and foreign vendors of reactor systems; the academic community; and the Electric Power Research Institute. (Information regarding the Committee's meetings is provided in Appendix A.)

The Committee agreed not to examine the desirability of further developing and deploying nuclear power. Instead, the Committee sought to answer the following question: If nuclear power is to be retained as an option for meeting U.S. electric energy requirements, what reactor technological options, associated Department of Energy research and development programs, and institutional changes would best serve that end? This study's purpose was not to advocate a new generation of nuclear power plants nor to assess the desirability of nuclear power relative to alternative energy sources. The Committee assumed that, at a later time, others would decide whether and under what conditions further development of nuclear power in the United States is warranted.

The Committee performed each of the tasks in its formal Statement of Task as follows:

1. The Committee identified the technological options for the next generation of reactors. In addition, after examining the range of reactor types that have active proponents, longer term options were identified (Chapter 3).
2. The Committee listed the criteria followed to address these options (Appendix B).
3. The Committee evaluated the options using these criteria as a framework (Chapter 3).
4. The Committee assessed the development effort required (Chapter 3) and prioritized what programs, in the Committee's judgment, should be funded (Chapter 4).
5. The Committee addressed which DOE facilities should be retained, depending on which alternative program is selected (Chapter 4).
6. The Committee addressed a variety of other aspects, primarily institutional, deemed necessary for retaining the nuclear option (Chapter 2).

7. The Committee developed policy alternatives in terms of near-term and long-term nuclear programs (Chapter 4).
8. The Committee provided its conclusions and recommendations (Chapter 5).

This report is the product of long hours of concentrated effort by the Committee members and the staff. The report benefited from the extensive comments provided by more than a dozen peer reviewers. Theresa Fisher devoted prodigious time to the many drafts. Norman Haller was extraordinary in his data searches and talent for accurate summarizing. His truly herculean efforts enabled this report finally to be produced. Archie Wood provided seasoned advice at critical times. Time and other constraints led several members of the original committee to resign. They nevertheless made important contributions, particularly Albert Babb.

John F. Ahearne
Chairman
Committee on Future Nuclear
Power Development

Table of Contents

Illustrations

FIGURES

TABLES

NUCLEAR POWER

Summary

From its beginnings in the early 1950s, the nuclear power industry and the government institutions that support and regulate it have brought nuclear generation to a position second only to coal as a source of electricity in the United States. By the end of 1990, the United States had 111 commercial nuclear power plants licensed to operate, with a combined capacity of about 99,000 megawatts electric.

However, expansion of commercial nuclear energy has virtually halted in the United States. No new nuclear plant has been ordered since 1978, scores of plants ordered earlier have been canceled, and construction of at least seven partially completed plants has been deferred. Concern for retaining an option for nuclear power led the Senate Appropriations Committee in 1988 to request the National Academy of Sciences to analyze the technological and institutional alternatives that would preserve the nuclear fission option in the United States.

The Committee on Future Nuclear Power Development was formed to conduct this study.

A premise of the Senate report directing this study is "that nuclear fission remains an important option for meeting our electric energy requirements and maintaining a balanced national energy policy." The Committee was not asked to examine this premise, and it did not do so. The Committee consisted of members with widely ranging views on the desirability of nuclear power. Nevertheless, all members approached the Committee's charge from the perspective of what would be necessary if we are to retain nuclear power as an option for meeting U.S. electric energy requirements, without attempting to achieve consensus on whether or not it should be retained. The Committee's conclusions and recommendations should be read in this context. The Committee's recommendations are identified by bold italicized type.

GENERAL CONCLUSIONS

The reasons that expansion of commercial nuclear energy has virtually halted in the United States include reduced growth in demand for electricity, high costs, regulatory uncertainty, and public opinion. Concern for safety, the economics of nuclear power, and waste disposal issues adversely affect the general acceptance of nuclear power.

Electricity Demand

Estimated growth in summer peak demand for electricity in the United States has fallen from the 1974 projection of more than 7 percent per year to a relatively steady level of about 2 percent per year. Ten year projections suggest a need for new capacity in the 1990s and beyond. To meet near-term anticipated demand, bidding by non-utility generators and energy efficiency providers is establishing a trend for utilities acquiring a substantial portion of this new generating capacity from others.

Nuclear power plants emit neither precursors to acid rain nor gases that contribute to global warming, like carbon dioxide. New regulations to address these environmental issues will lead to increases in the costs of electricity produced by combustion of coal, one of nuclear power's main competitors. Increased costs for coal-generated electricity will also benefit alternate energy sources that do not emit these pollutants.

Construction Costs and Times

Major deterrents for new U.S. nuclear plant orders include high capital carrying charges, driven by high construction costs and extended construction times, as well as the risk of not recovering all construction costs.

Data show a wide range of construction costs for U.S. nuclear plants, with the most expensive costing three times more (in dollars per kilowatt electric) than the least expensive in the same year of commercial operation. In the post-Three Mile Island era, the cost increases have been much larger. Considerable design modification and retrofitting to meet new regulations contributed to cost increases. The highest cost for a nuclear plant beginning commercial operation in the United States was twice as expensive (in constant dollars) from 1981 to 1984 as it was from 1977 to 1980. The average time to construct a U.S. nuclear plant went from about 5 years prior to 1975 to about 12 years from 1985 to 1989. U.S. construction times are much longer than those in most other major nuclear countries. Billions of dollars in disallowances of recovery of costs from utility ratepayers have made utilities and the financial community leery of further investments in nuclear power plants. During the 1980s, rate base disallowances by state regulators totaled about $14 billion for nuclear plants. Over the decade of the 1980s, operation and maintenance costs plus fuel costs for U.S. nuclear plants grew from nearly half to about the same as those for fossil fueled plants.

Performance

On average, U.S. nuclear plants have poorer capacity factors compared to those of plants in other Organization for Economic Cooperation and Development countries. On a lifetime basis, the United States is barely above 60 percent capacity factor, while France and Japan are at 68 percent, and West Germany is at 74 percent. U.S. plants averaged 65 percent in 1988, 63 percent in 1989, and 68 percent in 1990.

Except for capacity factors, the performance indicators of U.S. nuclear plants have improved significantly over the past several years. If the industry is to achieve parity with the operating performance in other countries, it must carefully examine its failure to achieve its own goal in this area and develop improved strategies, including better management practices. Such practices are important if the generators are to develop confidence that the new generation of plants can achieve the higher load factors estimated by the vendors.

Public Attitudes

Several factors seem to influence the public to have a less than positive attitude toward new nuclear plants: no perceived urgency for new capacity; nuclear power is believed to be more costly than alternatives; concerns that nuclear power is not safe enough; little trust in government or industry advocates of nuclear power; concerns about the health effects of low-level radiation; concerns that there is no safe way to dispose of high-level waste; and concerns about proliferation of nuclear weapons.

The following would improve public opinion of nuclear power:

- a recognized need for a greater electrical supply that can best be met by large plants;
- economic sanctions or public policies imposed to reduce fossil fuel burning;
- maintaining the safe operation of existing nuclear plants and informing the public;
- providing the opportunity for meaningful public participation in nuclear power issues, including generation planning, siting, and oversight;
- better communication on the risk of low-level radiation;
- resolving the high-level waste disposal issue; and
- assurance that a revival of nuclear power would not increase proliferation of nuclear weapons.

Safety

The risk to the health of the public from the operation of current reactors in the United States is very small. In this fundamental sense, current reactors are safe. However, a significant segment of the public has a different perception and also believes that the level of safety can and should be increased.

Institutional Changes

Large-scale deployment of new nuclear power plants will require significant changes by both industry and government.

Industry

One of the most important factors affecting the future of nuclear power in the United States is its cost in relation to alternatives and the recovery of these capital and operating charges through rates that are charged for the electricity produced. The industry must develop better methods for managing the design and construction of nuclear plants. Arrangements among the participants that would assure timely, economical, and high-quality construction of new nuclear plants will be prerequisites to an adequate degree of assurance of capital cost recovery from state regulatory authorities in advance of construction.

The financial community and the generators must both be satisfied that significant improvements can be achieved before new plants can be ordered. Greater confidence in the control of costs can be realized with plant designs that are more nearly complete before construction begins, plants that are easier to construct, use of better construction and management methods, and business arrangements among the participants that provide stronger incentives for cost-effective, timely completion of projects.

The principal participants in the nuclear industry--utilities, architect-engineers, and suppliers--should begin now to work out the full range of contractual arrangements for advanced nuclear power plants. Such arrangements would increase the confidence of state regulatory bodies and others that the principal participants in advanced nuclear power plant projects will be financially accountable for the quality, timeliness, and economy of their products and services.

Inadequate management practices have been identified at some U.S. utilities, large and small, public and private. A consistently higher level of demonstrated utility management practices is essential before the U.S. public's attitude about nuclear power is likely to improve. Over the past decade, utilities have steadily strengthened their ability to be responsible for the safety of their plants. Industry self-improvement, accountability, and self-regulation efforts improve the ability to retain nuclear power as an option for meeting U.S. electric energy requirements. The Committee encourages industry efforts to reduce reliance on the adversarial approach to issue resolution.

The nuclear industry should continue to take the initiative to bring the standards of every American nuclear plant up to those of the best plants in the United States and the world. Chronic poor performers should be identified publicly and should face the threat of insurance cancellations. Every U.S. nuclear utility should continue its full-fledged participation in the Institute of Nuclear Power Operations; any new operators should be required to become members through insurance prerequisites or other institutional mechanisms.

Standardization. A high degree of standardization will be very important for the retention of nuclear power as an option. There is not a uniformly accepted definition of standardization, although the industry has developed definitions of the various phases of standardization. A strong and sustained commitment by the principal participants will be required to realize the potential benefits of standardization (of families of plants) in the diverse U.S. economy. The following will be necessary:

- Families of standardized plants will be important for ensuring the highest levels of safety, realizing the potential economic benefits, and allowing standardized approaches to plant modification, maintenance, operation, and training.
- Customers must insist on standardization before an order is placed, during construction, and throughout the life of the plant.
- Suppliers must take standardization into account early in planning and marketing.
- Antitrust considerations will have to be taken into account.

Nuclear Regulatory Commission

An obstacle to continued nuclear power development has been the uncertainties in the Nuclear Regulatory Commission's (NRC) licensing process. Because the current regulatory framework was mainly intended for light water reactors (LWR) with active safety systems and because regulatory standards were developed piecemeal over many years, without review and consolidation, the regulations should be critically reviewed and modified (or replaced with a more coherent body of regulations) for advanced reactors of other types. ***The Committee recommends that NRC comprehensively review its***

regulations to prepare for advanced reactors, in particular, LWRs with passive safety features. The review should proceed from first principles to develop a coherent, consistent set of regulations.

NRC should improve the quality of its regulation of existing and future nuclear power plants, including tighter management controls over all of its interactions with licensees and consistency of regional activities. In addition, NRC should reduce reliance on the adversarial approach to issue resolution. ***The Committee recommends that NRC encourage industry self-improvement, accountability, and self-regulation initiatives.*** While federal regulation plays an important safety role, it must not be allowed to detract from or undermine the accountability of utilities and their line management organizations for the safety of their plants.

Economic incentive programs instituted by state regulatory bodies will continue for nuclear power plant operators. Properly formulated and administered, these programs should improve the economic performance of nuclear plants, and they may also enhance safety. However, they do have the potential to provide incentives counter to safety. Such programs should focus on economic incentives and avoid incentives that can directly affect plant safety. A joint industry/state study of economic incentive programs could help assure that such programs do not interfere with the safe operation of nuclear power plants. NRC should continue to exercise its federally mandated preemptive authority over the regulation of commercial nuclear power plant safety if the activities of state government agencies (or other public or private agencies) run counter to nuclear safety. Such activities would include those that individually or in the aggregate interfere with the ability of the organization with direct responsibility for nuclear plant safety (the organization licensed by NRC to operate the plant) to meet this responsibility. ***The Committee urges close industry-state cooperation in the safety area.***

The industry must have confidence in the stability of NRC's licensing process. Suppliers and utilities need assurance that licensing has become and will remain a manageable process that appropriately limits the late introduction of new issues.

It is likely that, if the possibility of a second hearing before a nuclear plant can be authorized to operate is to be reduced or eliminated, legislation will be necessary. The nuclear industry is convinced that such legislation will be required to increase utility and investor confidence to retain nuclear power as an option for meeting U.S. electric energy requirements. The Committee concurs.

Industry and the Nuclear Regulatory Commission

The U.S. system of nuclear regulation is inherently adversarial, but mitigation of unnecessary tension in the relations between NRC and its nuclear power licensees would, in the Committee's opinion, improve the regulatory environment and enhance public health and safety. Thus, the Committee commends the efforts by both NRC and the industry to work more cooperatively together and encourages both to continue and strengthen these efforts.

Department of Energy

Lack of resolution of the high-level waste problem jeopardizes future nuclear power development. The legal status of the Yucca Mountain site for a geologic repository should be resolved soon, and the Department of Energy's (DOE) program to investigate this site should be continued. A contingency plan must be developed to store high-level radioactive waste in surface storage facilities pending the availability of the geologic repository.

Environmental Protection Agency

Before operation of a high-level waste repository begins, DOE must demonstrate to NRC that the repository will perform to standards established by the Environmental Protection Agency (EPA). The EPA standard for disposal of high-level waste will have to be reevaluated to ensure that a standard that is both adequate and feasible is applied to the geologic waste repository.

Administration and Congress

The clear impression the Committee received from industry representatives was that protection such as the Price-Anderson Act would continue to be needed for advanced reactors, although some Committee members believe that this was an expression of desire rather than of need. At the very least, renewal of Price-Anderson in 2002 would be viewed by the industry as a supportive action by Congress and would eliminate the potential disruptive effect of developing alternative liability arrangements with the insurance industry.

Other

The Committee believes that the National Transportation Safety Board approach to safety investigations, as a substitute for the present NRC approach, has merit. In view of the infrequent nature of the activities of such a committee, it may be feasible for it to be established on an ad hoc basis and report directly to the NRC chairman. ***Therefore, the Committee recommends that such a small safety review entity be established.*** Before the establishment of such an activity, its charter should be carefully defined, along with a clear delineation of the classes of accidents it would investigate. Its location in the government and its reporting channels should also be specified.

Responsible arrangements must be negotiated between sponsors and economic regulators to provide reasonable assurances of complete cost recovery for nuclear power plant sponsors. Without such assurances, private investment capital is not likely to flow to this technology. Periodic reviews of construction progress and costs could remove much of the investor risk and uncertainty currently associated with state regulatory treatment of new power plant construction.

The institutional challenges are clearly substantial. If they are to be met, the Federal government must decide, as a matter of national policy, whether a strong and growing nuclear power program is vital to the economic, environmental, and strategic interests of the American people. Only with such a clearly stated policy, enunciated by the President and backed by the Congress through appropriate statutory changes and appropriations, will it be possible to effect the institutional changes necessary to return the flow of capital and human resources required to properly employ this technology.

Alternative Reactor Technologies

Advanced reactors are now in design or development. They are being designed to be simpler, and, if design goals are realized, these plants will be safer than existing reactors. The design requirements for the advanced reactors are more stringent than the NRC safety goal policy. An attractive feature of advanced reactors should be the significant reduction in system complexity and corresponding improvement in operability. While difficult to quantify, the benefit of improvements in the operator's ability to monitor the plant and respond to system degradations may well equal or exceed that of other proposed safety improvements.

The reactor concepts assessed by the Committee were the large evolutionary LWRs, the mid-sized LWRs with passive safety features,[1] the Canadian deuterium uranium (CANDU) heavy water reactor, the modular high-temperature gas-cooled reactor (MHTGR), the safe integral reactor (SIR), the process inherent ultimate safety (PIUS) reactor, and the liquid metal reactor (LMR). The Committee developed the following criteria for comparing these reactor concepts:

- safety in operation;
- economy of construction and operation;
- suitability for future deployment in the U.S. market;
- fuel cycle and environmental considerations;
- safeguards for resistance to diversion and sabotage;
- technology risk and development schedule; and
- amenability to efficient and predictable licensing.

Net Assessment

The Committee could not make any meaningful quantitative comparison of the relative safety of the various advanced reactor designs. The Committee believes that each of the concepts considered can be designed and operated to meet or closely approach the safety objectives currently proposed for future, advanced LWRs. The different advanced reactor designs employ different mixes of active and passive safety features. The Committee believes that there currently is no single optimal approach to improved safety. Dependence on passive safety features does not, of itself, ensure greater safety. The Committee believes that a prudent design course retains the historical defense-in-depth approach.

The economic projections are highly uncertain, first, because past experience suggests higher costs, longer construction times, and lower availabilities than projected and, second, because of different assumptions and levels of maturity among the designs. The Committee believes that the large evolutionary LWRs are likely to be the least costly to build and operate on a cost per kilowatt electric or kilowatt hour basis, while the high-temperature gas-cooled reactors and LMRs are likely to be the most expensive. The mid-sized LWRs with passive safety features lie between the two extremes.

SIR, MHTGR, PIUS, and LMR are not likely to be deployed for commercial use in the United States, at least within the next 20 years. The

---

[1] The term "passive safety features" refers to the use of gravity, natural circulation, and stored energy to provide essential safety functions in such LWRs.

development required for commercialization of any of these concepts is substantial.

It is the Committee's overall assessment that the large evolutionary LWRs and the mid-sized LWRs with passive safety features rank highest relative to the Committee's evaluation criteria. The evolutionary reactors could be ready for deployment by 2000, and the mid-sized could be ready for initial plant construction soon after 2000. The Committee's evaluations and overall assessment are summarized in Figure S-1.

The Committee has concluded the following:

1. Safety and cost are the most important characteristics for future nuclear power plants.

2. LWRs of the large evolutionary and the mid-sized advanced designs offer the best potential for competitive costs (in that order).

3. Safety benefits among all reactor types appear to be about equal at this stage in the design process. Safety must be achieved by attention to all failure modes and levels of design by a multiplicity of safety barriers and features. Consequently, in the absence of detailed engineering design and because of the lack of construction and operating experience with the actual concepts, vendor claims of safety superiority among conceptual designs cannot be substantiated.

4. LWRs can be deployed to meet electricity production needs for the first quarter of the next century.

a. The evolutionary LWRs are further developed and, because of international projects, are most complete in design. They are likely to be the first plants certified by NRC. They are expected to be the first of the advanced reactors available for commercial use and could operate in the 2000 to 2005 time frame. Compared to current reactors, significant improvements in safety appear likely. Compared to recently completed high-cost reactors, significant improvements also appear possible in cost if institutional barriers are resolved. While little or no federal funding is deemed necessary to complete the process, such funding could accelerate the process.

b. Because of the large size and capital investment of evolutionary reactors, utilities that might order nuclear plants may be reluctant to do so. If nuclear power plants are to be available to a broader range of potential U.S. generators, the development of the mid-sized plants with passive safety features is important. These reactors are progressing in their designs, through DOE and industry funding, toward certification in the 1995 to 2000 time frame. The Committee believes such funding will be necessary to complete the process. While a prototype in the traditional sense will not be required, federal funding will likely be required for the first mid-sized LWR with passive safety features to be ordered.

Reactor Designation	Available Design Information	Evaluation Criteria							
		Safety	Economy	Market Suitability	Fuel Cycle	Safeguards & Physical Sec.	Maturity of Development	Licensing	Overall Assessment[1]
ABWR[a]	○	○	○	○	◒	◒	○	○	○
APWR[a]	○	○	○	○	◒	◒	○	○	○
SYS 80+[a]	○	○	○	○	◒	◒	○	○	○
AP 600[b]	◒	○	◒	○	◒	◒	◒	○	○
SBWR[b]	◒	○	◒	○	◒	◒	◒	○	○
CANDU	○	○	●	◒	◒	◒	○	◒	◒
SIR	◒	○	◒[2]	●	◒	◒	●	●	●
MHTGR	◒	○	●	●	◒	◒	●	●	●
PIUS	●	○	◒[2]	●	◒	◒	●	●	●
PRISM[c] LMR	●	○	●[3]	●[3]	○	◒	●	●	◒

Legend Rating: ○ high ◒ moderate ● low

Notes: 1 Overall assessment was mostly driven by market suitability.
2 Lack of design maturity results in great uncertainty relative to vendor cost projections.
3 Long-term economy and market potential could be high, depending on uranium resource availability.
a ABWR - *advanced boiling water reactor;* APWR - *advanced pressurized water reactor;* and SYS 80+ - *system 80+. Large evolutionary LWRs.*
b AP 600 - *advanced pressurized 600;* and SBWR - *simplified boiling water reactor. Mid-sized LWRs with passive safety features.*
c PRISM - *power reactor, innovative small module.*

FIGURE S-1 Assessment of advanced reactor technologies.

This table is an attempt to summarize the Committee's qualitative rankings of selected reactor types ***against each other***, without reference either to an absolute standard or to the performance of any other energy resource options. This evaluation was based on the Committee's professional judgment.

c. Government incentives, in the form of shared funding or financial guarantees, would likely accelerate the next order for a light water plant. The Committee has not addressed what type of government assistance should be provided nor whether the first advanced light water plant should be a large evolutionary LWR or a mid-sized passive LWR.

5. The CANDU-3 reactor is relatively advanced in design but represents technology that has not been licensed in the United States. The Committee did not find compelling reasons for federal funding to the vendor to support the licensing.

6. SIR and PIUS, while offering potentially attractive safety features, are unlikely to be ready for commercial use until after 2010. This alone may limit their market potential. Funding priority for research on these reactor systems is considered by the Committee to be low.

7. MHTGRs also offer potential safety features and possible process heat applications that could be attractive in the market place. However, based on the extensive experience base with light water technology in the United States, the lack of success with commercial use of gas technology, the likely higher costs of this technology compared with the alternatives, and the substantial development costs that are still required before certification,[2] the Committee concluded that the MHTGR had a low market potential. The Committee considered the possibility that the MHTGR might be selected as the new tritium production reactor for defense purposes and noted the vendor association's estimated reduction in development costs for a commercial version of the MHTGR. However, the Committee concluded, for the reasons summarized above, that the commercial MHTGR should be given low priority for federal funding.

8. LMR technology also provides enhanced safety features, but its uniqueness lies in the potential for extending fuel resources through breeding. While the market potential is low in the near term (before the second quarter of the next century), it could be an important long-term technology, especially if it can be demonstrated to be economic. The Committee believes that the LMR should have the highest priority for long-term nuclear technology development.

9. The problems of proliferation and physical security posed by the various technologies are different and require continued attention. Special attention will need to be paid to the LMR.

---

[2] The Gas Cooled Reactor Associates estimates that, if the MHTGR is selected as the new tritium production reactor, development costs for a commercial MHTGR could be reduced from about $1 billion to $0.3 - 0.6 billion.[DOE, 1990 in Chapter 3]

Alternative Research and Development Programs

The Committee developed three alternative research and development (R&D) programs, each of which contains three common research elements: (1) reactor research using federal facilities. The experimental breeder reactor-II, hot fuel examination facility/south, and fuel manufacturing facility are retained for the LMR; (2) university research programs; and (3) improved performance and life extension programs for existing U.S. nuclear power plants.

The Committee concluded that federal support for development of a commercial version of the MHTGR should be a low priority. However, the fundamental design strategy of the MHTGR is based upon the integrity of the fuel ($\leq$1600°C) under operation and accident conditions. There are other potentially significant uses for such fuel, in particular, space propulsion. Consequently, the Committee believes that DOE should consider maintaining a coated fuel particle research program within that part of DOE focused on space reactors.

Alternative 1 adds funding to assist development of the mid-sized LWRs with passive safety features. Alternative 2 adds a LMR development program and associated facilities--the transient reactor test facility, the zero power physics reactor, the Energy Technology Engineering Center, and either the hot fuel examination facility/north in Idaho or the Hanford hot fuel examination facility. This alternative would also include limited research to examine the feasibility of recycling actinides from LWR spent fuel, utilizing the LMR. Finally, Alternative 3 adds the fast flux test facility and increases LMR funding to accelerate reactor and integral fast reactor fuel cycle development and examination of actinide recycle of LWR spent fuel.

None of the three alternatives contain funding for development of the MHTGR, SIR, PIUS, or CANDU-3.

Significant analysis and research is required to assess both the technical and economic feasibility of recycling actinides from LWR spent fuel. The Committee notes that a study of separations technology and transmutation systems was initiated in 1991 by DOE through the National Research Council's Board on Radioactive Waste Management.

It is the Committee's judgment that Alternative 2 should be followed because it:

- provides adequate support for the most promising near-term reactor technologies;
- provides sufficient support for LMR development to maintain the technical capabilities of the LMR R&D community;

- would support deployment of LMRs to breed fuel by the second quarter of the next century should that be needed; and
- would maintain a research program in support of both existing and advanced reactors.

1

Introduction

From its beginnings in the early 1950s, the nuclear power industry[1] and the government institutions that support and regulate it have brought nuclear generation to a position second only to coal as a source of electricity in the United States. By the end of 1990, the United States had 111 commercial nuclear power plants licensed to operate[2], with a combined capacity of about 99,000 megawatts electric.[NRC, 1991] In 1989 nuclear plants produced about 19 percent of the nation's electric power: 529 billion kilowatt hours, much more energy than nuclear power provided in France and Japan combined.[IAEA, 1990] Three more U.S. plants are now under construction.[3][NRC, 1991] U.S. nuclear power technology has provided the basis for nuclear power plants worldwide.[Gavrilas et al., 1990] In 1989, nuclear plants produced 77 percent of France's electricity, 26 percent of Japan's electricity, and 33 percent of West Germany's electricity.[DOE, 1991]

However, expansion of commercial nuclear energy has virtually halted in the United States. No new nuclear plant has been ordered since 1978, scores of plants ordered earlier have been canceled, and construction of at least seven partially completed plants has been deferred. In other countries, too, growth of nuclear generation has slowed or stopped.

---

[1] Terms such as "the nuclear power industry," "the nuclear industry," and "the industry" are used throughout this report. In the broadest sense, these terms include the utilities that operate nuclear plants; the architect-engineers, nuclear steam supply system vendors, and other suppliers that help the utilities design and construct nuclear plants or develop and manufacture the nuclear and non-nuclear components that are installed in nuclear plants to generate electricity; and the various organizations that support these entities (e.g., "industry"-sponsored organizations that perform research or interface with regulatory agencies on nuclear matters).

[2] This number excludes the Rancho Seco plant in California that was shut down as a result of a referendum vote in mid 1989 (see Chapter 2). It also excludes the Shoreham plant in New York that was shut down before receiving a full power license.

[3] Watts Bar Units 1 and 2 in Tennessee, and Comanche Peak Unit 2 in Texas.

The reasons for this interruption in growth are varied. In the United States, growth in demand for electricity has slowed from about 7 percent annually in the early 1970s to about 2 percent today. In addition, extended construction schedules and high costs of building nuclear power plants have been of great concern, and the costs of operating and maintaining nuclear power plants have risen more rapidly than those of a principal competitor--coal plants.[DOE, 1988] The cost of base load generated electricity is strongly affected by capital costs, and therefore it is relatively sensitive to factors such as inflation, high interest rates, delays, and backfit requirements that increase the cost of construction. This is particularly so for nuclear plants, compared to electricity from fossil plants that are more sensitive to inflation in fuel costs. Also, state public utility commissions have disallowed billions of dollars of construction costs from inclusion in rate bases (and thus from recovery from utility customers). These disallowances have made utility executives and the financial community leery of further investments in nuclear power. Public concerns about reactor safety, fed by the accidents at Three Mile Island in 1979 and Chernobyl in 1986, have led to organized local and statewide opposition to new nuclear power plants. Finally, the federal government's failure to meet schedules in assuring the safe disposition of spent reactor fuel has further tarnished nuclear energy in public opinion.

In the 1980s, reactor vendors in the United States and other countries initiated development of new reactor technologies with features intended to provide lower cost construction and operation, improved reactor safety, and in some cases greater flexibility in adding capacity. This report assesses these designs and outlines several alternative research and development programs that would ready new nuclear power technology for use in the future. It also addresses issues of future electricity demand, cost, utility management, public opinion, safety, and licensing and regulation that bear on the future of nuclear power.

THE COMMITTEE'S CHARGE

In requesting this study, Congress asked the National Academy of Sciences to analyze the technological and institutional alternatives that would preserve the nuclear fission option in the United States. The Senate Appropriations Committee report accompanying the 1989 Energy and Water Development Appropriation bill said:

> [The Senate Committee on Appropriations] believes that nuclear fission remains an important option for meeting our electric energy requirements and maintaining a balanced national energy policy. The Committee continues to strongly support the need for a responsive nuclear fission program, but finds the current civilian nuclear power reactor program to be a deficient aggregate of numerous reactor

> types and conceptual variations being developed without the guidance of well-defined strategic objectives. The Committee finds further that the future development and institutionalization of nuclear power development should be rethought, newly defined, and directed to be responsive to current and projected conditions. Therefore, the Committee specifically provides . . . for a critical comparative analysis by the National Academy of Sciences of the practical technological and institutional options for future nuclear power development and for the formulation of coherent policy alternatives to guide the Nation's nuclear power development.[U.S. Congress, 1988]

The Committee on Future Nuclear Power Development was formed to conduct this study.

The Committee consisted of members with widely ranging views on the desirability of nuclear power. Nevertheless, all members approached the Committee's charge from the perspective of what would be necessary if we are to retain nuclear power as an option for meeting U.S. electric energy requirements, without attempting to achieve consensus on whether or not it should be retained. The Committee's conclusions and recommendations should be read in this context.

REFERENCES

DOE, Energy Information Administration. 1991. International Energy Annual. DOE/EIA-0219(89). February.

DOE, Energy Information Administration. 1988. An Analysis of Nuclear Power Plant Operating Costs. DOE/EIA-0511. Released for printing March 16, 1988.

Gavrilas, M., P. Hejzlar, Y. Shatilla, and N. Todreas. 1990. Report on Safety Characteristics of Light Water Reactors of Western Design. Department of Nuclear Engineering, Massachusetts Institute of Technology. Cambridge, MA. Preliminary (in publication). December 12, 1990.

IAEA. 1990. Nuclear Power Reactors in the World. Reference Data Series Number 2. Vienna, Austria. April.

NRC. 1991. U.S. Nuclear Regulatory Commission Information Digest, 1991 Edition. NUREG-1350. 3(March).

U.S. Congress, Senate Committee on Appropriations. 1988. Calendar No. 726. Sen. Rep. 100-381, Energy and Water Development Appropriation Bill, 1989. Report to accompany H.R. 4567. Ordered to be printed June 9, 1988.

2

The Institutional Framework

This study examines the key institutional issues that have affected U.S. nuclear power development for the past 20 years. These issues will also strongly shape nuclear power's future and must be adequately accommodated to retain nuclear power as an option for meeting U.S. electric energy requirements.

The major issues that are examined here are not new--they have been widely recognized and discussed since at least the early 1980s. For example, one study in 1983 tried to identify what it is that prevents nuclear power from going forward in the United States by looking at "The Utility Director's Dilemma."

> . . . What is the risk to the company that after it invests $2-3 billion in a 12- to 14-year process of constructing a new nuclear power plant, the plant will not be able to operate? What is the risk to the utility that the return on the $2-3 billion invested will be zero? What is the risk that events beyond the control of the company, and beyond its analysts' best forecasts, will delay by several years the date on which the plant comes on line, will double the cost, or will otherwise affect its operation in a manner that could destroy the stockholders' equity and the utility?
>
> If the decision to order the new nuclear power plant were made today [i.e., in 1983], the plant could begin producing power between 1995 and 1997. What could happen in the interim? Could some future president or Congress, governor or state legislature, Nuclear Regulatory Commission (NRC) or public utilities commission (PUC) be antinuclear? . . . Is there reasonable likelihood of an accident of Three Mile Island (TMI) proportions or worse during the ensuing 12-14 years at one or more of the 200 nuclear power plants operating in the world? How could that affect public opinion, political referenda, and, thus, the prospects for the utility's new nuclear plant?[Allison and Carnesale, 1983]

In the mid-1980s the Congressional Office of Technology Assessment found, after a major study, that

> Without significant changes in the technology, management, and level of public acceptance, nuclear power in the United States is unlikely to be expanded in this century beyond the reactors already under construction. Currently, nuclear power plants present too many financial risks as a result of uncertainties in electric demand growth, very high capital costs, operating problems, increasing regulatory requirements, and growing public opposition.
>
> If all these risks were inherent to nuclear power, there would be little concern over its demise. However, enough utilities have built nuclear reactors within acceptable cost limits, and operated them safely and reliably to demonstrate that the difficulties with this technology are not insurmountable.[U.S. Congress, 1984]

At about the same time, a study by researchers at the Massachusetts Institute of Technology reached the following conclusion.

> Despite the best efforts at institutional reform and innovation in LWR [light water reactor] technology, the difficulties presently confronting the U.S. nuclear power industry are sufficiently serious and persistent that the utilities may not overcome their present unwillingness to order new LWRs during the 1990s, even if faced with a need to build large amounts of new central station baseload capacity at that time. [Lester et al., 1985]

One senior electric utility executive put it another way in 1985.

> Apart from everything else, expansion of the nuclear power option in the United States is not likely to occur unless and until there is broad public and political support for it.[Willrich, 1985]

In 1989, another study examined the question, "Will nuclear power recover in a greenhouse?" It contained the following summary:

> The major problems in the United States which led to removing nuclear power as a choice for new generating capacity were lack of growing demand for electricity, rising costs per plant, and bad management, as well as growing public opposition. Unless these issues are recognized and addressed, greenhouse warming will not lead to nuclear power being chosen when utility executives select technologies to pursue for meeting new demands. Actions by Congress, the public, and the industry are needed.[Ahearne, 1989]

The issues addressed by this Committee stem in large part from observations such as those cited above as well as from personal experience.[1] Often interrelated in complicated ways, these issues include future electricity demand and supply, costs (and disallowances of costs), utility management, public opinion, safety, waste management, proliferation, and licensing and regulation.

FUTURE ELECTRICITY GENERATION

Future Demand

Estimated growth in summer peak demand for electricity in the United States has fallen from the 1974 projection of more than 7 percent per year to a relatively steady level of about 2 percent per year. Table 2-1 shows the projected average annual rates of summer peak demand growth over various 10-year periods, according to the North American Electric Reliability Council. The table also shows actual average annual growth rates in summer peak demand. The data indicate that projections made in the mid-to-late 1970s were too high. Enough time has not passed to know whether projections made in the 1980s will be correct.

For the period 1990 to 1999, the North American Electric Reliability Council projects that summer peak demand will increase from about 539,000 megawatts electric (MWe) to about 646,000 MWe, an average annual growth rate of 2.0 percent per year. The Council estimates that there is an 80 percent probability that the actual average annual growth over the period will not exceed 2.7 percent per year or fall below 1.2 percent per year.[North American Electric Reliability Council, 1990]

---

[1] There were, of course, other studies not mentioned here. See, for example, Nuclear Power in America, by William Lanouette [Lanouette, 1985], the Report of the Edison Electric Institute on Nuclear Power [EEI Task Force on Nuclear Power, 1985], An Acceptable Future Nuclear Energy System, Condensed Workshop Proceedings [Firebaugh et al., 1980], the Energy Research Advisory Board's Report to the Department of Energy, Review of the Proposed Strategic National Plan for Civilian Nuclear Reactor Development [DOE, 1986a], and other references in this report.

TABLE 2-1 Projected and Actual Summer Peak Demand Growth Rates by Year of the Estimate

Year of the Estimate	Ten-Year Average Annual Projected Growth Rates (percent)	Actual Average Annual Growth Rates (percent)
1974	7.6	2.9 (through 1983)
1978	5.2	2.3 (through 1987)
1982	3.0	3.5 (through 1990)
1986	2.2	*3.5 (through 1990)
1988	1.9	
1990	2.0	

*NOTE: This is only 4 years of data.

SOURCES: [U.S. Congress, 1984; North American Electric Reliability Council, 1991, 1990, 1989, 1988, 1987, and 1986; DOE, 1986c]

The Energy Information Administration has prepared long-range estimates of growth in U.S. electricity demand. The Energy Information Administration also compared its estimates to four other forecasts. Table 2-2 summarizes the results, which range from a low of 1.6 percent per year to a high of 2.6 percent per year average annual growth from 1988 to 2010.[2]

Finally, the Edison Electric Institute (EEI) has prepared a forecast to the year 2015. The EEI estimates an average annual growth rate in electricity demand of about 2.6 percent per year for 1987 to 2000, dropping to 1.5 percent per year for 2000 to 2015.[EEI, 1989]

Future Supply

In 1989, the United States had an installed summer generating capacity of about 673,000 MWe. During the 1990 to 1999 period, the North American Electric Reliability Council estimates U.S. additions of about 86,000 MWe and retirements of about 4,000 MWe. Average projected annual growth in installed generating capacity equals about 8,000 MWe per year. The Council

[2] DOE's National Energy Strategy, published in February 1991, provides the following growth rate projections for U.S. electricity consumption under the National Energy Strategy Scenario: 1990 to 2000 - 2.5 percent per year; 2000 to 2010 - 1.5 percent per year; 2010 to 2020 - 1.6 percent per year; and 2020 to 2030 - 1.3 percent per year.[DOE, 1991]

TABLE 2-2 Projections of Growth in U.S. Electricity Demand, 1988 to 2010

Source of Forecast	Average Annual Growth Rates (percent)
Energy Information Administration	2.1 to 2.6[a]
Gas Research Institute	2.0
American Gas Association	1.9
WEFA Group	1.9
DRI/McGraw Hill	1.6

[a] The Energy Information Administration's Base Case forecast is 2.3 percent. Ranges extend from 2.1 percent per year to 2.6 percent per year depending on assumptions about oil prices and economic growth rates.

SOURCE: [DOE, 1990a]

indicates that, in 1999, total U.S. installed summer and winter generating capacity will be about 761,000 MWe and 779,000 MWe, respectively.[North American Electric Reliability Council, 1990]

Long-range forecasting has many uncertainties. Nevertheless, beyond the year 1999, a plausible scenario for supply growth rates might lie between 1.5 and 2.6 percent per year, the long-range demand forecasts given earlier. Starting from the larger estimated winter value of 779,000 MWe at the end of 1999, such growth rates would then produce supply growths of about 12,000 MWe per year to 20,000 MWe per year. If retirements of, for example, 1,000 MWe per year were assumed, new additions would need to be about 13,000 MWe per year to 21,000 MWe per year for the first several years of the next century.[3]

---

[3] During the 1990s, the North American Electric Reliability Council estimated that the largest number of U.S. retirements would be about 700 MWe in the year 1996.[North American Electric Reliability Council, 1990] The use of such figures, especially after the year 1999, assumes that aging, clean air standards, or strong pressures to reduce carbon dioxide generation do not force large scale retirement of nuclear or fossil plants. Significantly larger numbers of retirements could, of course, directly affect the need for new capacity. For example, if the licenses of currently operating nuclear plants are not extended, nuclear retirements would be about 6,000 MWe per year during the period 2005 to 2010.[NRC, 1991a]

The new capacity would consist of both baseload and peaking units, which would be provided both by traditional utility rate-based sources and by "non-utility generators," or independent power producers and companies with generating facilities that qualify under the Public Utility Regulatory Policies Act of 1978 (PURPA). Such facilities would include cogeneration and small hydroelectric plants, for example.

Finally, some additional supply capacity is likely to be satisfied by a combination of further energy-efficiency improvements, renewable energy technologies, gas, coal, and repowering.[4] Thus, the annual need for new nuclear capacity, at least during the first several years of the next century, is likely to be only a portion of the new additions (which are estimated to be 13,000 MWe to 21,000 MWe per year). This prospect is in contrast to that of the peak years of nuclear plant orders when, from 1970 to 1974, new orders for nuclear units averaged about 31,000 MWe per year [DOE, 1989a], although many of these were later cancelled.[5]

Growth in Competition

Due to high facility development and construction costs and state regulatory practices, utilities today are more often contracting with third party power producers through competitive bidding procedures designed to acquire new generating capacity.[6] According to a recent national survey, since 1984,

---

[4] Accompanying a warning of electricity shortages in this decade, the report of a recent conference stated "A full mix of options and enough lead time to make sound choices on both demand and supply sides is far safer than short-term decisions and catch-up policies. Choices need to reflect local, regional and global environmental priorities, as well as the economics and reliability of the entire electric supply and delivery system."[Fowler and Rossin, 1990]

[5] The Atlantic Council of the United States indicated that no nuclear power plants that have been ordered since 1973 have been put into construction for the simple reason that "about twice as many units were on order as were needed with the abrupt decline in the rate of growth of electric power demand. . . . "[Atlantic Council of the United States, 1990]

[6] One review of responses to bidding requests for proposals indicates that, in 16 states, responses exceeded requests by a factor of 8 (38,674 megawatts in response to requests for 4,781 megawatts).[Blair, 1990]

27 states have adopted or are developing competitive procurement systems that, together with access already granted by PURPA, will affect the nation's electric power markets.[7] [National Independent Energy Producers, 1990] Experience so far suggests that a substantial portion of new generating capacity can be purchased in this fashion.[DOE, 1989b]

> Because several years are often required to construct generating sources, utilities have little operating experience with competitively purchased electricity. Thus, the effects of competitive power purchases on the long-term reliability of electric service--which is affected by the reliability of all sources and transmission and distribution facilities--are not yet certain and difficult to assess.[U.S. General Accounting Office, 1990]

According to the electricity supply estimates for 1990 through 1999 made by the North American Electric Reliability Council, about 18,000 MWe of non-utility generator additions are planned compared to about 68,000 MWe of utility generating unit additions.[North American Electric Reliability Council, 1990] In 1990, 6,000 MWe of non-utility generation went into service, bringing the total to 32,700 MWe.[National Independent Energy Producers, 1991]

---

[7] The Congressional findings underlying PURPA are " . . . that the protection of the public health, safety, and welfare, the preservation of national security, and the proper exercise of congressional authority under the Constitution to regulate interstate commerce require--

(1) a program providing for increased conservation of electric energy, increased efficiency in the use of facilities and resources by electric utilities, and equitable retail rates for electric consumers;

(2) a program to improve the wholesale distribution of electric energy, the reliability of electric service, the procedures concerning consideration of wholesale rate applications before the Federal Energy Regulatory Commission, the participation of the public in matters before the Commission, and to provide other measures with respect to the regulation of the wholesale sale of electric energy;

(3) a program to provide for the expeditious development of hydroelectric potential at existing small dams to provide needed hydroelectric power;

(4) a program for the conservation of natural gas while insuring that rates to natural gas consumers are equitable;

(5) a program to encourage the development of crude oil transportation systems; and

(6) the establishment of certain other authorities as provided in title VI of this Act."[U.S. Congress, 1978]

Others estimate the likely non-utility share at 50 percent or more.[National Independent Energy Producers, 1990]

The entities currently entering the independent power production bidding process are offering cost-competitive generating plants that use well-established gas-fired or renewable generating technologies with short construction lead times. In general, fixed-price contracts are used for construction. These circumstances do not now favor large-scale baseload technologies.

Integrated Resource Planning

The goal of integrated resource planning is to minimize the societal costs of the reliable energy services needed to sustain a healthy economy. Many utilities have installed or are installing new planning systems to assure that all options to supply electricity are considered and the least-cost options are chosen.[National Association of Regulatory Utility Commissioners, 1988; EPRI, 1988] Untapped electricity savings from end-use efficiency improvements are treated explicitly as a resource option, functionally comparable to energy deliveries to consumers from power plants. Comparisons among resource options are made on the basis of life cycle costs, and efforts are often made to incorporate environmental costs in some fashion.[Cohen et al., 1990]

These systems usually make the planning process more open and more competitive. Such systems have been pioneered in California and in the Pacific Northwest under the aegis of the California Energy Commission and the Northwest Power Planning Council. Integrated resource planning activities are also under way in many other states, including Arizona, Illinois, Maryland, Nevada, New York, Wisconsin, and the New England States. The National Association of Regulatory Utility Commissioners has formally endorsed this planning concept. These systems are intended to ensure that energy-efficiency improvements and supply-side technologies of all types, including future nuclear power generation, are compared on an equal basis. It remains to be seen whether these systems will favor, be neutral toward, or be negative regarding nuclear power.

Environmental Factors

Nuclear power plants emit neither precursors to acid rain nor gases that contribute to global warming, like carbon dioxide. Both of these environmental issues are currently of great concern. New regulations to address these issues will lead to increases in the costs of electricity produced by combustion of coal, one of nuclear power's main competitors.

Technology is already available to limit emissions of sulfur and nitrogen oxides from coal-based plants, the principal acid rain precursors, and new technology is being developed in the clean coal technology program at the Department of Energy (DOE) and elsewhere. However, even with this new technology, emissions of these pollutants will be much greater than those associated with the nuclear cycle. These technologies will add to the cost of electricity generated in coal-fired plants and will affect the future competition between coal and nuclear plants. Increased costs for coal-generated electricity will also benefit alternate energy sources that do not emit these pollutants.

No practical way to capture and contain carbon dioxide emissions is now available. Depending on the growth in concern about global climate change, controls on the combustion of fossil fuels to reduce such emissions could severely limit the use of coal, oil, and to a lesser extent natural gas-fired generation and could make nuclear power more attractive. Energy efficiency and renewable generating technologies would realize similar benefits.

ELECTRICITY GENERATION COSTS

In order to deliver electricity, it must first be generated, then transported and distributed to individual users. This report considers only the costs of electricity generation, which consist of the sum of capital carrying charges, operation and maintenance costs, and fuel costs. Capital carrying charges are, in essence, the cost of capital and the depreciation and amortization of the costs of building and financing the plant.[8] Such charges are the predominant cost of generating electricity with nuclear power. Furthermore, capital carrying costs are constantly changing as additional investments are required over the life of the plant.

In this section, each of the components of costs of nuclear generated electricity is examined in order to understand its importance. Construction times for nuclear plants are discussed as well because of their significance to capital carrying charges. Some cost comparisons with coal are also presented. International data are provided where appropriate.

---

[8] See Electric Plant Cost and Power Production Expenses 1988 [DOE, 1990c] for a more complete discussion of the costs included in capital carrying charges. Decommissioning costs can also be included.[DOE, 1982; Jones and Woite, 1990] Operation and maintenance expenses and fuel expenses will be defined later.

Capital Carrying Charges

The Energy Information Administration (EIA) analyzed the cost components in 1988 for major U.S. privately owned nuclear and coal-fired plants. There is wide variation among the highest and lowest total generation costs and among the components of that cost, as seen in Table 2-3. For example, nuclear plants have both the lowest and highest total generation costs in the table. The difference between the high and low ends is due almost entirely to large differences in the capital carrying charges (approximately a factor of 20 for both nuclear and coal). On the average, the data show that nuclear plant capital carrying charges are about three times that of coal plants, accounting for the major net difference between their total generation expenses.[DOE, 1990c]

TABLE 2-3 Components of Highest, Lowest, and Average Total Generating Costs in 1988 for Nuclear and Coal-Fired Plants Owned by Major Private Utilities (Cents per Kilowatt Hour)[a]

	Highest Total Generation Costs		Lowest Total Generation Costs		Average Total Generation Costs	
	Nuclear	Coal	Nuclear	Coal	Nuclear	Coal
Total Costs[b]	11.3	8.5	1.6	2.2	5.6	3.1
Components[c]						
Capital Carrying	9.4	5.4	0.4	0.3	3.4	1.1
Operation & Maintenance	1.2	0.7	0.8	0.5	1.5	0.4
Fuel	0.7	2.5	0.5	1.4	0.8	1.7

There were 179 major privately owned electric utilities in the United States in 1988. Specific definition of the term "major" is contained in the report entitled Electric Plant Cost and Power Production Expenses 1988.[DOE, 1990c]

[a] These data can be interpreted as the price of electricity generated in 1988 from nuclear and coal-fired plants and do not represent the cost of producing electricity over the entire life of the plants.

[b] Numbers may not add due to rounding

[c] In the first four columns, these are the costs for each component for the plants whose total costs were highest and lowest. The last two columns represent the average plant (e.g., the average total nuclear costs are 5.6, made up of 3.4, 1.5, and 0.8).

SOURCE: [DOE, 1990c]

Existing nuclear plants have higher capital carrying charges, on average, for several reasons: (1) their equipment and buildings have been more expensive to acquire than those for coal-fired plants, (2) they have taken longer to build, thus accumulating more interest during construction (which is, in many cases, capitalized), and (3) in most cases, nuclear plant decommissioning costs are taken into account in the capital carrying charges. These are all reflected in capital carrying charges in Table 2-3.

The large amounts of capital required to build and finance some U.S. nuclear power plants are a major cause of disenchantment with the technology. The Committee was unable to find a complete and consistent set of data on such costs. Therefore, to analyze the fundamental reasons for large differences in the costs among U.S. nuclear plants, the Committee makes use of the best data found.

One measure of capital investment is called ***historical plant cost***.[9] Another measure is construction cost in ***mixed-current dollars***.[10] Although such measures mix dollars over many years, they do suggest that both nuclear and large (≥300 MWe) fossil-fueled plants have exhibited cost increases over time. For example, large fossil-fueled plants that entered commercial service from 1976 to 1978 had average historical costs of about $300 per kilowatt electric, whereas those entering commercial service in 1987 had average historical costs of about $1,000 per kilowatt electric. Nuclear units beginning commercial operation from 1976 to 1978 had average construction costs (mixed-current dollars) of about $600 per kilowatt electric, whereas those beginning commercial operation in 1987 had average construction costs of about $4,000 per kilowatt electric.[DOE, 1990c and 1989d]

Because historical plant cost and mixed-current dollar construction cost data are difficult to use for cost comparisons, analysts have devised ways of

---

[9] Historical plant costs are the net cumulative-to-date actual outlays or expenditures for a facility. These costs are effectively those that enter the rate base and are recovered from ratepayers. Historical costs contain dollar values of the year in which the expenditure occurred; thus they are a mixture of dollars in different time periods. Differences in accounting practices also affect such costs, for example, the inclusion or exclusion of time-related costs such as allowance for funds used during construction (AFUDC). For more explanation see the report entitled Electric Plant Cost and Power Production Expenses 1988.[DOE, 1990c]

[10] These costs are referred to variously as final reported completion costs and final estimates of total construction cost for nuclear units. The costs are in current dollars of a number of different years (e.g., expenditures in 1971 are in 1971 dollars).[DOE, 1989d]

separating out the time-related costs and converting the resulting costs to constant dollars. Such costs are called ***overnight costs*** (i.e., the cost of a plant if it could be built instantaneously, or overnight).[11] It is difficult to produce such numbers, but they are important. As will be seen in Chapter 3, one element of the prospective merit of new nuclear plants is their predicted overnight capital cost.

Table 2-4 summarizes the overnight construction costs in constant 1988 dollars that have been calculated for 76 U.S. nuclear power plants. The table shows that there is a wide range between the highest and lowest values of overnight cost for each time period. From 1977 to 1988, for example, the highest cost plants were three times as expensive as the lowest cost plants entering commercial operation in the same time periods. Furthermore, the data show continued escalation in overnight costs for plants beginning commercial operation during the 1970s and 1980s, with a sharp increase from the years before 1981 to 1981 and beyond (e.g., the highest cost plant from 1981 to 1984 was twice as expensive as the highest cost plant from 1977 to 1980). The cost increases do not appear to be affected strongly by the introduction of larger plants.[12]

The higher standards of quality and quality assurance required for nuclear plants were not initially sufficiently appreciated by the nuclear industry nor its regulators. This lack of appreciation contributed in many cases to inadequate quality, and even occasionally to mistakes, in construction, with attendant higher costs.

Time-related costs (i.e., those costs that result because the nuclear plant cannot be constructed overnight, such as financing charges) accounted for approximately 25 percent of the inflation-adjusted increase in total construction costs.[DOE, 1986b] Thus, the time-related costs are significant,

---

[11] For more discussion, see the reports entitled An Analysis of Nuclear Power Plant Construction Costs [DOE, 1986b] and The Economics of Nuclear Power, Further Evidence on Learning, Economies of Scale, and Regulatory Effects.[Cantor and Hewlett, 1988]

[12] However, one study points out that, because of the indirect effect of size on costs, there seems to be some evidence supporting claims that attempts were made to build plants too large to be efficiently managed by the constructors.[Cantor and Hewlett, 1988]

TABLE 2-4 Overnight Construction Costs for Selected U.S. Nuclear Power Plants, by Year of Commercial Operation [b]

Year of Commercial Operation	1988 Dollars per kWe[a]			Number of Plants	Average Size (MWe)
	Highest	Lowest	Average		
1971-1974	1,234	480	817	13	855
1975-1976	1,284	562	1,035	12	885
1977-1980	1,608	562	1,118	11	905
1981-1984	3,326	1,003	1,733	15	1,064
1985-1986	4,204	1,342	2,620	15	1,129
1987-1988	4,596	1,383	3,133	10	1,070

SOURCE: [DOE, 1986b], supplemented by revised overnight cost data base provided by Energy Information Administration staff on December 14, 1990 (See Note below)

[a] 1982 Dollars in source material were converted to 1988 Dollars by using factor of 1.213 [DOE, 1989c]

[b] Nuclear Regulatory Commission data were used for dates of commercial operation and for individual plant capacities [NRC, 1990a,b]

NOTE: The data base that was provided contained 79 plants. Seabrook, Shoreham, and Three Mile Island 2 were excluded from the above calculations because the Seabrook costs only went to 1986, the Shoreham plant never reached commercial operation and costs only went to 1985, and Three Mile Island 2, which had $1.173 billion in overnight costs (1988 dollars) through 1978, was destroyed in early 1979. Plants not included in the data base were turnkey plants (for which the reported costs were not believed representative of the realized costs) and plants for which data were not available. The procedure used to compute the overnight costs consisted of starting with the historical plant costs (i.e., those that entered the rate base), and then removing the time related costs (i.e., interest and inflation). The results were the actual cash outlays for construction. The accounting procedures used by the utilities for reporting these cash outlays are governed by the Uniform System of Accounts (a set of federal regulations). Thus, the accounting variations that remain are very small.[J. Hewlett, Energy Information Administration, personal communication]

although not as significant as the overnight costs.[13] For this reason, it is useful to examine the time it takes to construct nuclear power plants.

Construction times are an important issue because long construction times increase capital carrying charges, directly through finance charges and indirectly through growth in the costs of labor and materials. Long construction times also increase the possibility of regulatory changes that may require expensive plant modifications.

Table 2-5 shows nuclear power plant construction times for 110 U.S. plants entering commercial service through 1989. The table indicates continually growing construction times from when the first U.S. nuclear plants entered commercial service through the 1980s. The minimum times doubled from about 3 years to 6-7 years, and the maximum times went from about 8 years to 13-19 years. The average times grew from 5 years to 10-12 years.

Table 2-6 shows how U.S. construction times compare to those experienced by other countries, particularly France (see the article entitled Nuclear Units Under Construction [Bacher and Chapron, 1989] for discussion of French nuclear plant construction), West Germany, Japan, and the United Kingdom. For units beginning operation prior to 1978, all countries took about the same time to construct a nuclear plant--4 to 6 years on the average. Afterwards, however, the United Kingdom's and the United States' average times doubled (to about 11 to 13 years). There were also increases in the construction times experienced by France,[14] West Germany, and Japan, but they were not so pronounced.

The high costs of recently completed nuclear plants have been subjected to intense review by state public utility commissions, and in some cases

---

[13] See An Analysis of Nuclear Power Plant Construction Costs [DOE, 1986b] for a more complete discussion of the relative importance of overnight costs versus time-related costs. In particular, that report contains the following statement: "In short, in real, inflation-adjusted terms, escalation in overnight costs, rather than time-related costs, is the principal factor causing the cost increases. Thus, attempts to reduce costs should focus on the managerial and regulatory factors that affect plant design and construction, as well as on the factors that just affect the time required for licensing of the plants."

[14] The increase of the construction times in France is mainly due to the fact that the power of the French pressurized water reactors was raised in successive steps, from 900 MWe initially to 1,300 MWe and lately to 1,450 MWe.

TABLE 2-5 Construction Times For 110 U.S. Nuclear Power Plants. Construction Times in Years[a] for Plants Beginning Commercial Operation in Given Time Periods

	Prior to 1975	1975 through 1979	1980 through 1984	1985 through 1989
Minimum Time	2.7	3.7	6.1	6.6
Maximum Time	7.6	10.1	13.4	19.3[b]
Average Time	5.4	7.2	10.1	12.2
Number of Plants	40	23	17	30

[a] Construction Time is defined here as time elapsed from actual ground breaking until the first generation of electrical energy.
[b] This plant first generated electricity in September 1989, even though it did not begin commercial operation until 1/8/90.

SOURCE: [NRC, 1982 and 1990f]

utilities have not been allowed to fully recapture the capital costs of plants in rates. The aggregate value of these disallowances and the reasons for them will be discussed after the next section. Mid-construction cancellations have created an additional source of financial risk. Between 1972 and 1984, more than $20 billion in capital flowed into 115 nuclear projects that their sponsors later cancelled.[Cavanagh, 1986] When a plant is cancelled, some costs are recovered by the utility from customers and others are not, depending on rulings of the applicable regulatory commissions.

TABLE 2-6 Comparison of U.S. Nuclear Power Plant Construction Time Spans with Those of Other Countries. Average Construction Time Spans in Years[a] for Plants[b] Connected to Grids in Given Time Periods

Country	Prior to 1978	1978 through 1989
France	5.1	5.9
West Germany	4.7	7.6[c]
Japan	3.9	4.7
United Kingdom	5.7	12.8[d]
United States	5.5	11.1[e]
World (including United States)	5.2	7.7

[a] Time spans measured from first pouring of concrete to unit connection with grid.
[b] Both operating and shut down reactors are included.
[c] However, the average time span of all (four) nuclear power plants for which construction began in West Germany after the Three Mile Island accident was 5.7 years.
[d] These were gas reactors.
[e] This includes about a two year regulatory delay after the Three Mile Island accident.

SOURCE: [IAEA, 1990]

Major deterrents to new orders for nuclear plants include their high capital carrying charges, which are driven by high construction costs and extended construction times, and the risk that their construction costs will not be recovered. Both of these issues (i.e., reduced capital carrying charges and predictability of cost recovery) must be addressed before new nuclear plant orders are likely.

Operation, Maintenance, and Fuel Costs

Rising costs of the operation and maintenance (O&M) of a nuclear power plant after it has been constructed are also an important consideration in the decision to build a new plant. These O&M expenses, as they are sometimes called, are defined as follows:

> Operation expenses are associated with operating a facility (i.e., supervising and engineering expenses). Maintenance expenses are that portion of expenses consisting of labor, materials, and other direct and indirect expenses incurred for preserving the operating

> efficiency or physical condition of utility plants that are used for power production, transmission and distribution of energy.[DOE, 1990c]

Fuel costs are the last component of total electricity generation costs. Fuel costs

> . . . include the fuel used in the production of steam or for driving another prime mover for the generation of electricity. Other associated expenses include unloading the shipped fuel and all handling of the fuel up to the point where it enters the first tank, bunker, hopper, bucket, or holder in the boiler-house structure.[15][DOE, 1990c]

Comparative trends from 1982 through 1988 for the average O&M costs and fuel costs of both nuclear and fossil-fueled plants are displayed in Table 2-7. These data show that nuclear O&M costs have increased significantly through the 1980s, while fossil fuel costs have decreased significantly.[16,17] This result has led EIA to state

> The advantage seen for nuclear power in fuel cost is diminished by their operation and maintenance expenses and high capital costs. [DOE, 1990c]

Previously, after completing a detailed analysis of the trend of O&M costs for nuclear power plants, EIA stated

> Continued escalation in operating costs could erode any cost advantage that operating nuclear power plants now have. . . . If

---

[15] Apparently, waste disposal costs are not included in the fuel costs of DOE's Electric Plant Cost and Power Production Expenses 1988 report [DOE, 1990c] for either coal or nuclear plants.

[16] At this time, of course, concerns about oil prices and effects of the Clear Air Act amendments raise questions about the stability of fossil fuel prices.

[17] In an earlier report EIA provided data that, when adjusted to constant 1988 cents, showed that the sum of O&M and fuel costs for coal plants was nearly 3 cents per kilowatt hour versus nearly 2 cents per kilowatt hour for nuclear plants. However, by 1987, the sum was identical (a little more than 2 cents per kilowatt hour) for both coal and nuclear plants.[DOE, 1989e]

TABLE 2-7 Average Operation and Maintenance and Fuel Costs for Nuclear and Fossil-Fueled[a] Plants Owned by Major Private Utilities[b] (1988 cents per kilowatt hour)[c]

Year	Operation and Maintenance		Fuel		O&M plus Fuel[d]	
	Nuclear	Fossil-Fueled	Nuclear	Fossil-Fueled	Nuclear	Fossil-Fueled
1982	1.1	0.5	0.7	3.0	1.8	3.5
1983	1.2	0.5	0.8	2.8	1.9	3.3
1984	1.3	0.5	0.8	2.7	2.1	3.2
1985	1.2	0.5	0.8	2.6	2.0	3.1
1986	1.3	0.5	0.8	2.1	2.1	2.6
1987	1.4	0.5	0.8	1.9	2.2	2.4
1988	1.5	0.5	0.8	1.8	2.2	2.3

[a] In 1988 coal accounted for almost 80 percent of the generation from fossil-fueled steam electric plants owned by the major private electric utilities, gas 11 percent, petroleum 9 percent, and a small percentage from wood and waste.
[b] There were 179 major privately owned electric utilities in the United States in 1988. Specific definition of the term "major"si contained in DOE's Electric Plant Cost and Power Production Expenses 1988 report.[DOE, 1990c]
[c] Inflation indices used for 1982 to 1988 through 1987 to 1988 were 1.21, 1.17, 1.13, 1.09, 1.07, and 1.03, respectively. These indices were obtained from the implicit price deflators contained in DOE's Annual Energy Review 1989.[DOE, 1989c]
[d] Numbers may not add horizontally due to rounding

SOURCE: [DOE, 1990c]

operating costs continue to escalate, it may become economical to close some of the older plants, and thus the assumption of a 40-year operating life may be optimistic.[DOE, 1988c]

Three important factors were given by the EIA for the changes in nuclear O&M costs. These factors appear below.

Over the period studied, increases in the price of replacement power (i.e., power from other sources to replace the power lost when a nuclear power plant is out of service) offered an increased incentive to improve performance, resulting in increased O&M costs. Furthermore, State regulatory actions provided additional incentives to improve plant performance. In total, these economic and regulatory incentives to improve plant performance statistically explained about 15 percent of the escalation in real O&M costs. The analysis could find no evidence that increases in replacement power prices influenced real capital additions costs.

Plant aging has received a great deal of attention, and some analysts have cited aging as a major determinant of nuclear power plant operating costs. However, this analysis found that plant aging explained only about 17 percent of the escalation in capital additions costs. Furthermore, as plants age, real O&M costs actually fell . . . this could be due to the fact that as plants age, the experience of the operator increases, which could result in lower costs: with all other factors held constant, real O&M costs would be about 33 percent less because of plant aging.

A third important factor is the effect of increases over time in NRC activity. Unfortunately, this analysis was unable to separate these NRC regulatory effects from those resulting from increases in industry experience and any other unmeasurable factor correlated with time. However, the combined effects of all these time-related factors were substantial. In the absence of increases in NRC regulatory activity, industry learning, and other unmeasurable factors correlated with time, real O&M and capital additions costs would be about 70 percent and 60 percent lower, respectively, than otherwise.[DOE, 1988c]

Nuclear Costs in Other Countries

U.S. nuclear power plants represent only about a fourth of those in the world.[IAEA, 1990] Therefore, it is instructive to examine the relative cost competitiveness of nuclear and coal plants through the perspectives of other nations. It is also instructive to review the means that others have used to control the costs of nuclear power plants.

The results of examining the *projected* costs of nuclear power plants versus coal plants were summarized in a recent journal article based on several studies, including ones by the Organization of Economic Cooperation and Development and the International Union of Producers and Distributors of Electrical Energy.[Jones and Woite, 1990; Moynet et al., 1988; OECD, 1989] The authors did caution against the comparison of absolute costs between countries.[18] However, some large differences indicate the influence of institutional aspects that might have considerable importance for the economics of nuclear energy. The low investment costs of French plants to a large extent may be due to more efficient use of engineering capacity by construction of larger series of similar plants (standardization) and by the concept of twin and multiple units. The recent German experience with the Konvoi plants (decrease of investment costs in German currency) also indicates the importance of standardization.

Relative to the United States, reported O&M costs are much lower in most other countries. For France, values of 0.5 cent per kilowatt hour are reported for 1987, which is about one third of the current U.S. average. Influencing factors may be the more efficient organization of engineering infrastructure within one large utility, smaller work force, and the better employment of work force in multiple unit plants. For West Germany some increase in O&M costs has been observed, which is due to safety improvements of operating plants (backfits and accident management implementation). Still, the West-German O&M costs are about half the U.S. average.

Nevertheless, the journal article does present data that provide an assessment of the *projected* relative costs of nuclear and coal plants as well as the electricity they could generate. These data show that the investment cost of coal-fired plants was projected to be less than that of nuclear power plants on a dollars per kilowatt electric basis for all 22 countries examined, including the United States. For some countries (e.g., France and Czechoslovakia), projected nuclear investment costs exceeded those of coal by only a small amount. For other countries (e.g., Germany and China), nuclear plants were projected to require twice the investment of coal plants.

The electricity generation costs of projected coal plants versus nuclear power plants were also compared on a relative basis. The results were summarized by the authors as follows:

---

[18] Reasons are " . . . because of their substantial variations of economic and social systems and different provisions for radioactive waste management, plant decommissioning, and environmental protection."[Jones and Woite, 1990]

In short, most of the participating countries expect nuclear power to have a lower levelized generating cost than coal-fired generation or, at worst, to about break-even. However, for most countries, the projected comparisons between coal and nuclear generating costs are not clear cut when viewed across the full range of assumptions considered in the studies. Under some assumptions of parameter values, nuclear power has a sizeable cost advantage over coal; for other parameter values the reverse is the case.[Jones and Woite, 1990]

With respect to bringing the benefits achieved in the most successful countries to others, the authors tabulated ways of reducing the capital costs of nuclear power plants (e.g., feedback experience through standardization, extend planning quality and quantity by completing detailed designs and resolving political and regulatory issues before starting construction, and improve project management).[Jones and Woite, 1990]

Costs of Disallowances

U.S. nuclear plants were designed and built using a variety of arrangements utilizing architect-engineers, equipment suppliers, constructors, consultants, and internal staffs. Many of these arrangements resulted in timely construction, within budget. Others experienced significant delays and serious cost overruns.

These overruns attracted the attention of state public utility commissions. Many "prudency" reviews[19] followed. These reviews resulted in disallowances of cost recovery and have become a major risk of the construction of nuclear power plants. As Table 2-8 and Table 2-9 show, these disallowances have been for a variety of reasons, including capacity not needed and imprudence

---

[19] These are post-construction reviews to determine which costs of construction were prudently incurred and should be recovered from ratepayers.

TABLE 2-8 Nuclear and Non-Nuclear Disallowances by Issue During the 1980s (Number of cases or disallowances are given in parentheses)

Category	Regulatory Issue[a]	$ in Billions Nuclear	Non-Nuclear
Capital	Excess Capacity	$ 2.2 (7)	$0.6 (7)
	Economic Value	1.0 (4)	0
	Imprudence	7.1 (27)	0.1 (5)
	Cost Caps	2.7 (9)	0
	Other	0.8 (9)	0
Subtotal (Capital)		$13.8 (44)[b]	$0.7 (12)
Operations	Imprudence	$ 0.6 (44)	$ [c]
Totals		$14.4 (88)	$0.7 (12)

[a] The definitions of each type of regulatory issue are provided in the note below.
[b] Some cases involve capital disallowances on more than one issue.
[c] Non-nuclear operating imprudence not examined.

SOURCE: EEI Rate Regulation Department, December 1990 and January 1991

NOTE to Table 2-8:

- Excess Capacity - investments disallowed as not "used and useful" to the public. Such disallowances are not necessarily permanent and can be included in the rate base at a later date, if the utility's capacity requirements have grown sufficiently. Some of this capacity may be sold to other utilities needing such capacity.
- Economic Value - investments disallowed as excessive in comparison to alternate sources of generation. The difference between book and market value usually was excluded from rate base.
- Imprudence - capital costs said to have been imprudently incurred. Typically, imprudence findings have centered on decisions which affected the schedule for completion of the plant, or which involved the management of engineering and construction tasks.
- Cost Caps - investments disallowed in order not to exceed a predetermined cap on the rate base value of a project. In most cases, such caps were the result of negotiated settlements.
- Other - disallowed investments that do not fall within the above categories (e.g., rate case settlements that cannot be attributed to a specific issue described above).

Note to Table 2-8 (continued)

- Nuclear Operating Imprudence - nuclear operating expenses said to have been imprudently incurred. In these decisions, PUCs typically disallowed incremental replacement power costs or maintenance expenditures incurred as a result of imprudently extended outages (i.e., actions, or the lack thereof, which unreasonably increased the length of outages).

(e.g., alleged mismanagement).[20] The financial losses from these disallowances have been borne by utility stockholders. The prospect of the inability to recover all costs of nuclear construction through rates is a major deterrent to future investments in nuclear power. Without reasonable assurance of cost recovery, private utilities will have difficulty in obtaining new equity as well as debt capital to help finance any baseload generation.

The state public utility commissions have demonstrated that incurred costs that the commissions have deemed imprudent will not be recovered. This is an authority of utility commissions that has seldom been used before. It has been primarily applied to nuclear, rather than fossil, power plants.[21] Knowing that costs might be considered "prudent" or "imprudent," the industry must develop better methods for managing the design and construction of nuclear plants. Arrangements among the participants that would assure timely, economical, and high-quality construction of new nuclear plants, the Committee believes, will be prerequisites to an adequate degree of assurance of capital cost recovery from state regulatory authorities in advance of construction. The development of state prudency laws also can provide a positive response to this issue (see discussion later in this chapter).

---

[20] Another source indicated that the disallowances for imprudent actions were $13 billion during the period 1984 to 1988. "That represents an average disallowance of 14 percent of all plants judged to have imprudent actions associated with it, including both alleged mismanagement of construction and excess capacity judgments."[Cohen, 1989]

[21] Table 2-8 demonstrates that disallowance for non-nuclear plants represented a small fraction of the total.

TABLE 2-9 Nuclear Disallowances By Year During the 1980s (Number of cases or disallowances are given in parentheses)

	$ in Billions										
Category	1980	1981	1982	1983	1984	1985	1986	1987	1988	1989	Totals
Capital	0	0	0	0.3(1)	0.1(1)	4.2(8)	3.7(14)	1.5(6)	1.8(7)	2.2(7)	13.8(44)
Operations[a]	- (1)	- (3)	0.1 (4)	- (2)	- (3)	0.1(12)	0.1(5)	0 (3)	- (6)	0.1(5)	0.6[b](44)

[a] "-" means less than $50 million
[b] Numbers do not add because of rounding

SOURCE: EEI (Edison Electric Institute) Rate Regulation Department, December 1990

Liability Protection

The nuclear industries have been covered since 1957 by the Price-Anderson Act. This act limits the maximum liability of the nuclear industry to a catastrophic accident. This limit is now about $7 billion (i.e., $200 million in primary liability insurance plus $63 million per plant for 100+ plants if the primary insurance is exceeded). In case of an accident, money would be collected by insurance pools from all nuclear plant operators and paid to claimants on behalf of the plant that had the accident. No more than $10 million per plant per year would be collected.[Presidential Commission on Catastrophic Nuclear Accidents, 1990] Damage costs above this amount would probably, but not necessarily, fall to the federal government to pay; in any event, federal payments would require legislation by the Congress. The Price-Anderson Act was renewed in 1988 and will expire in 2002 unless it is renewed again by the Congress.[Price-Anderson Amendments Act of 1988]

In its original consideration of this legislation, Congress had an estimate by the Atomic Energy Commission of the possibilities and consequences of severe nuclear power accidents. (This estimate was entitled "Theoretical Possibilities and Consequences of Major Accidents in Large Nuclear Power Plants," WASH 740, March 1957.) Since then, the estimated probabilities and consequences of major large accidents have changed. The question arises: Will this liability limitation still be needed as nuclear industry protection after 2002, or can the nuclear industry rely upon its own resources? The clear impression the Committee received from nuclear industry representatives was that such protection would continue to be needed, although some Committee members believe that this was an expression of desire rather than of need. At the very least, renewal of Price-Anderson in 2002 would be viewed by the industry as a supportive action by Congress and would eliminate the potential disruptive effect of developing alternative liability arrangements with the insurance industry. Failure to renew Price-Anderson in 2002 would raise a new impediment to nuclear power plant orders as well as possibly reduce an assured source of funds to accident victims.

UTILITY MANAGEMENT OF CONSTRUCTION AND OPERATIONS

Currently, 53 utilities are licensed to operate nuclear plants in the United States[NRC, 1991a] The federal government made an early commitment to nuclear power. Plant construction was initiated based on limited research, development, and demonstration. Many reactor suppliers and many architect-engineers and contractors launched ambitious plans to secure market share. This hindered the sharing of experiences nationwide as well as the development of efficiencies usually associated with a learning curve. As costs escalated, fixed price contracting often shifted to cost plus arrangements, with a consequent reduction of control over costs by utilities.

Concurrent Design and Construction

As seen in Table 2-10, construction of large nuclear plants began before lessons could be learned from operating the early smaller nuclear plants. Additionally, construction of most plants began with incomplete designs,[22] a practice that proved to be a problem for this then-emerging technology. Such problems were exacerbated by regulatory standards that were developed piecemeal over many years, without review and consolidation, as issues arose in the construction and operation of current-generation plants.[23]

These regulations have been criticized by the industry as redundant, confusing, and in some instances, contradictory. Because this regulatory framework evolved with and is mainly intended for LWRs with active safety systems, it should be critically reviewed and modified (or replaced with a more coherent body of regulations) for advanced reactors of other types, particularly LWRs incorporating passive safety systems.

In 1989 NRC issued 10 CFR Part 52, the new licensing rule, which for certification requires that designs of evolutionary LWRs be "essentially complete." For certification of other types of reactors, 10 CFR Part 52 requires that there be either (1) " . . . analysis, appropriate test programs, experience, or a combination thereof; . . . sufficient data . . . on the safety features . . . ;" and a design that is "complete" in "scope," or (2) " . . . acceptable testing of an appropriately sited, full-size, prototype of the design. . . ."[24] [GSA, 1990] A strict application of such requirements would assure both that design concurrent with construction would be minimized and that future

---

[22] "Construction on many recent nuclear plants was begun with <15 percent of the plant design complete."[Chung and Hazelrigg, 1989]

[23] In addition to the basic rules, those in the Code of Federal Regulations (10 CFR--), NRC publishes regulatory guides, branch technical positions, and assorted other advisories. As of 1987, the 10 CFR regulations filled more than 1,000 pages of the federal code, and NRC had 141 regulatory guides on power reactors and reactor-related areas.[Ahearne, 1988]

[24] See Chapter 3 for more discussion of certification requirements for reactors other than evolutionary LWRs. In November 1990, the U.S. Court of Appeals overturned a portion of 10 CFR Part 52, but left the remainder, including this requirement, untouched. (See later discussion in the Licensing and Regulation Section.)

TABLE 2-10 Comparison of Average Sizes of U.S. Nuclear Power Plants in Commercial Operation with U.S. Plants Receiving Construction Permits

Year	Average Size In Commercial Operation (MWe)	Average Size Construction Permit (MWe)	Ratio (Op'n/Constr'n)
1964	118	542	0.2
1965	118	610	0.2
1966	118	704	0.2
1967	118	747	0.2
1968	310	801	0.4
1969	412	909	0.5
1970	466	877	0.5
1971	540	946	0.6
1972	574	871	0.7
1973	624	1,052	0.6
1974	686	1,064	0.6
1975	713	1,151	0.6
1976	729	1,148	0.6
1977	752	1,027	0.7
1978	756	860	0.9

SOURCE: [NRC, 1989a]

nuclear plants would have a high degree of standardization. The Electric Power Research Institute's Advanced Light Water Reactor Utility Requirements Document specifies that designs be 90 percent complete prior to beginning construction (see Chapter 3, Table 3-2).

Standardization

There were five major competing suppliers of first-generation nuclear reactors in the United States. Three of these were suppliers of pressurized LWRs of similar basic design, one supplied boiling LWRs, and one, high-temperature gas-cooled reactors.

LWR vendors competed by offering reactors of ever-increasing capacity to take advantage of expected economies of scale. In addition, most of the utilities tailored designs of individual plants to their own special requirements. Rather than adhering to a single design and slowly and systematically improving it, suppliers, architect-engineers, and utilities made substantial changes to each design. Additionally, virtually every plant was modified extensively throughout the construction phase. The result was that most U.S. plants became customized, unique designs.

There is general belief that standardization of nuclear plants will result in accelerated licensing, improved construction schedules, lower capital costs, and increased safety. There is little evidence in the United States, however, to verify this claim.

In the past, changes in regulatory requirements and individual utility preferences made licensing of replicate plants nearly impossible. Claims for improved quality and cost control for factory built modules, as compared to on-site fabrication, have not yet been substantiated.

Other countries have developed their nuclear power industries differently. In France, for example, there is a single reactor vendor and architect engineer, and a single utility. France concentrated on a single technology, the pressurized water reactor design, and exploited this technology with uniform construction practices and evolutionary design upgrades in a disciplined, controlled process.[Giraud and Vendryes, 1989]

Furthermore, Electricité de France has capitalized on its standardized designs by standardizing many aspects of maintenance, operations, and training. Many observers believe that the vigorous approach to standardization, in both design and operation, has been an important factor in the overall success of the French program.

Of the nuclear stations in the United States, only a very few are "standardized." (However, in a number of cases, 2, and in some instances 3, nearly identical units are located at the same site.) Examples include the two SNUPPS units, Calloway and Wolf Creek; the four units at Byron and Braidwood; and the three units at Palo Verde. The operators of these units point to substantial benefits from a limited approach to standardization.

Achieving a significant degree of standardization will prove to be very challenging. The new NRC licensing rule provides a vehicle for encouraging standardization. However, achieving a high level of standardization through 10 CFR Part 52 is likely to be expensive and time consuming. The debate over implementation of Part 52 has revealed that there is not a uniformly accepted definition of standardization. Also, utility industry efforts to create and sustain common patterns of purchasing behavior may raise anti-trust concerns that will need careful review (the Committee did not address this issue).

The industry, under the auspices of the Nuclear Power Oversight Committee (NPOC), has developed a position paper on standardization that provides definitions of the various phases of standardization and expresses an industry commitment to standardization. The NPOC paper discusses four phases: (1) Standardization of Utility Requirements, (2) Standardization of Design Certification and Standardized Licensing, (3) Commercial Standardization,[25] balance of plant," and (4) Standardization Beyond Design.[Nuclear Power Oversight Committee, 1991]

The Committee believes that a strong and sustained commitment by the industry's principal participants (utilities, suppliers, and architect-engineers) will be required to realize the potential benefits of standardization (of families of plants) in the diverse U.S. economy.

Plant Management

Many nuclear plants in the United States have operated very well over extended periods of time. Their managements have been identified as being

---

[25] To illustrate the definitions provided by NPOC, the Committee has extracted the following example:

> Commercial standardization expands the level of design standardization achieved under design certification...in that it addresses design decisions beyond regulatory requirements and provides design standardization outside the regulatory scope.
>
> Commercial standardization is the nonrecurring engineering which can be performed generically and applied directly to all plants referencing the same design certification. Simply stated, commercial standardization begins with the level of design detail required for design certification and concludes with the level of design detail where site-specific and project-specific characteristics control. Since the level of detail required for design certification will vary based on the safety significance of the system, it follows that the starting point for commercial standardization will also vary by system. Commercial standardization will also vary by system. Commercial standardization includes all of the engineering needed to complete the nonrecurring engineering tasks for a family of plants. It will include procurement, construction, and installation specification details beyond those required for design certification, including function, fit, and form details for standardized equipment. Prior to beginning construction, some recurring engineering must be completed to account for site-specific and project-specific items. Site-specific differences are minimized by employing a 'site-envelope' design approach that bounds most U.S. sites; therefore, site differences should not significantly reduce the degree of standardization.

competent and responsible. Effective plant management has been identified both by NRC and by the industry as critical to safe, economic operation of nuclear power plants. Inadequate management practices can have serious consequences, as noted in the examples below:

- Pilgrim, shut down by NRC for over four years because of poor management;
- Peach Bottom, shut down for two years after operators were found asleep in the control room, with strong criticism of management made by the Institute of Nuclear Power Operations (INPO) and NRC;
- Rancho Seco, closed by referendum after an extended period of inadequate management by both the board of directors and line management;
- Tennessee Valley Authority's (TVA) nuclear program, one of the largest in the United States, shut down completely for more than four years; and
- Washington Public Power Supply System, about $8 billion to produce 1 operating plant, 2 moth-balled plants, and 2 cancelled plants.

Such management failures increase skepticism about and opposition to nuclear power generally. Today, the whole utility industry therefore has a stake in helping to improve the management practices of its weakest members, or as a last resort, to insist that the weakest members not operate nuclear power plants.[26] Because of the high visibility of nuclear power and the responsibility for public safety, a consistently higher level of demonstrated utility management practices is essential before the U.S. public's attitude about nuclear power is likely to improve.[27]

---

[26] One industry report analyzed the industry's problems and recommended many initiatives. The report stated that some utilities are not living up to appropriate standards, and recommended that the industry publicly identify such utilities.[Nuclear Power Oversight Committee, 1986]

[27] Emphasis upon current reactor operations has been stressed by Norman Rasmussen:[George Washington University, 1989]

> First and foremost ... we must operate today's reactors safely and efficiently for the next 5 to 10 years, and create a climate where people begin to accept reactors as good neighbors that produce electricity rather cheaply and don't pollute the environment. We don't need any more Pilgrims, any more Peach Bottoms, or any more problems like TVA.
>
> ... that is mainly an industry responsibility to get more serious about it, put in the proper management, and run more reactors the way we've already demonstrated that 25 percent of them run.

U.S. nuclear utilities made an industry-wide commitment to address collectively their management problems seriously through establishment of INPO following the TMI accident.[28] INPO has become highly regarded. Its establishment was clearly a major step toward improving nuclear operations and toward better communication among nuclear plant operators. Trends in the performance indicators discussed below confirm that operation of many plants is improving, although as shown previously O&M costs are also rising rapidly.

Plant Performance

On average, U.S. nuclear power plants have not achieved a capacity factor (or load factor)[29] as high as planned, or as high as is obtained in many other countries. This gives some hope that the cost of nuclear power can be reduced by proper attention to plant performance. In this section we present a historical survey of the trends of load factors and other performance indicators.

For plants devoted to baseload operation, as most nuclear plants are, load factor is a good indicator of performance. As load factors increase, plants produce more electrical energy in a given time period. The International Atomic Energy Agency collects load factor data for the world's nuclear power plants. A ten year overview of load factors for countries that are members of the Organization for Economic Cooperation and Development (OECD) is provided in Table 2-11. This table indicates that, from 1978 through 1987, U.S. nuclear plants had an average load factor of about 60 percent, with the highest annual average (68 percent) occurring in 1978 and the lowest annual average (58 percent) occurring in 1983. The U.S. lifetime average also was

---

[28] The 1979 report of the President's Commission on the Accident at Three Mile Island contained a recommendation that the nuclear industry must "set and police its own standards of excellence." In response, INPO was formed.[INPO, 1989] The industry also formed the Nuclear Safety Analysis Center, the Non-Destructive Evaluation Center, and the Nuclear Maintenance Assistance Center to deal with specific aspects of nuclear plant safety and performance improvements.

[29] In the United States capacity factor is the ratio (expressed as a percentage) of actual electrical energy generation to the electrical energy that could have been generated if a unit ran continuously at maximum capacity during a given time period. The International Atomic Energy Agency defines load factor in essentially the same way. For the purposes of this section, capacity factor and load factor will be used interchangeably.

TABLE 2-11 Load Factors of Nuclear Power Plants (OECD Countries)

Country	1978	1979	1980	1981	1982	1983	1984	1985	1986	1987	Lifetime
Belgium	81.5	74.2	81.3	83.9	83.3	79.4	86.8	82.9	77.8	82.7	81.2
Canada[a]	76.9	78.3	85.1	89.9	87.4	86.5	76.2	70.9	72.9	72.4	78.2
Finland	79.0	82.4	58.7	77.7	84.9	86.0	88.8	89.2	89.0	91.6	85.1
France[b]	72.4	57.8	67.4	65.2	57.4	65.9	74.5	74.7	71.2	64.7	68.1
Germany FR	63.3	55.5	54.9	67.3	71.1	71.5	82.4	85.7	77.9	78.7	73.6
Italy[c]	92.0	31.0	25.0	23.3	59.5	43.4	57.9	53.7	74.6	1.6	47.1
Japan	54.3	49.2	61.4	61.2	70.1	69.9	72.1	74.1	76.3	79.3	68.0
Netherlands	87.8	74.4	91.5	77.8	83.7	77.0	77.1	82.4	90.3	74.5	81.5
Spain	77.6	68.0	52.8	68.2	47.3	48.4	65.9	64.8	74.1	80.6	67.6
Sweden[b]	70.2	62.1	70.4	69.5	66.0	64.9	75.4	75.3	79.6	76.3	71.7
Switzerland	89.3	88.1	80.0	84.6	84.4	86.8	89.4	84.2	83.3	84.2	85.0
United Kingdom[d]	62.7	62.6	59.0	59.4	69.7	78.9	79.4	82.6	69.4	66.0	69.1
United States	68.0	60.9	58.4	60.5	58.6	57.5	58.4	61.4	59.1	60.2	60.5

[a] Pressurized heavy water reactors
[b] Affected in later years by load following
[c] Political moratorium in 1987
[d] Gas-cooled reactors

NOTE:

- Only non-prototype reactors ≥ 100 MWe considered.
- All permanently shut-down reactors are excluded.
- Load factor calculated from the month following the date of commercial operation.

SOURCE: [OECD, 1989 and IAEA PRIS, Report NBLG020G 89-02-02]

about 60 percent as of 1987. On a lifetime basis, only Italy had poorer performance, due at least in part to the political moratorium on nuclear power in 1987.[OECD, 1989] In 1988 the U.S. capacity factor was 65 percent, in 1989, 63 percent, and in 1990, 68 percent.[NRC, 1991a]

Some U.S. nuclear plants perform very well compared to others in the world, while some do very poorly. For example, the list of 22 top performing plants in OECD nations (those plants with lifetime load factors to 1988 of 85 percent or more) contains only 3 U.S. plants. On the other hand, of the 22 bottom performing plants (lifetime load factors to 1988 under 50 percent), there are 12 U.S. plants.[IAEA, 1990]

There have been claims by some U.S. utilities that special surveillance, backfit, and maintenance requirements specified by NRC extend normal refueling outage times beyond that for fuel changeout only. An annual outage duration of 2 weeks limits the maximum possible capacity factor to 96 percent, and 10 weeks limits the maximum to 80 percent. Accordingly, a number of utilities have changed from annual refueling to extended operating schedules employing 18-month or 24-month refueling schedules. If a 6-week outage is scheduled every 12 months, the maximum capacity factor possible is 88 percent; if every 24 months, 94 percent. Attempts to improve U.S. nuclear plant capacity factors by scheduling less frequent refueling outages are receiving increasing attention. The Committee did not quantify differences between outage durations in the United States and those in other countries attributable to regulatory requirements.

Other nuclear power plant performance indicators of interest are the number of unplanned automatic reactor scrams (i.e., trips or shutdowns) while a reactor is critical, the number of selected safety system actuations, and the collective radiation exposure per plant. INPO publishes such data for U.S. plants, and they appear in Figure 2-1.[30] The data show considerable improvements in the industry averages over the 1980s. The 1990 goals shown in Figure 2-1 were established in 1985.

---

[30] The Federal Republic of Germany has some similar data.[Birkhofer, 1991]

Equivalent availability factor

Equivalent availability factor is the ratio of the total power a unit could have produced, considering equipment and regulatory limits, to its rated capacity, expressed as a percentage.

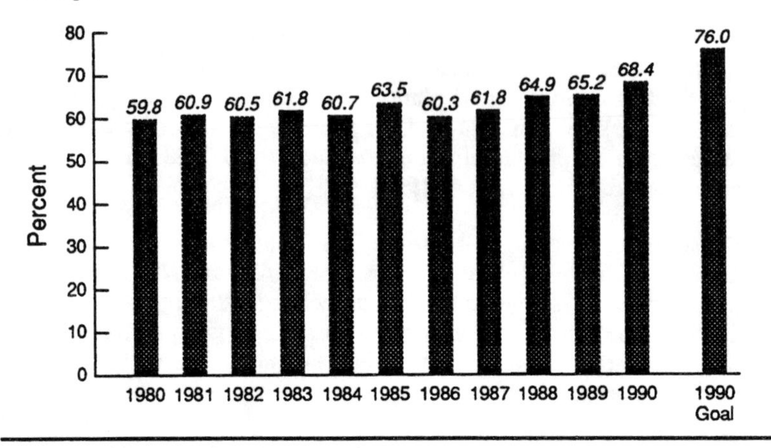

Unplanned automatic scrams

The graph shows the average number of unplanned automatic scrams while the reactor is critical that occurred at nuclear plants operating with an annual capacity factor of 25 percent or greater.

NRC's "Automatic Scrams While Critical" indicator and the INPO indicator differ in the criteria for including new units and units operated for part of a year.

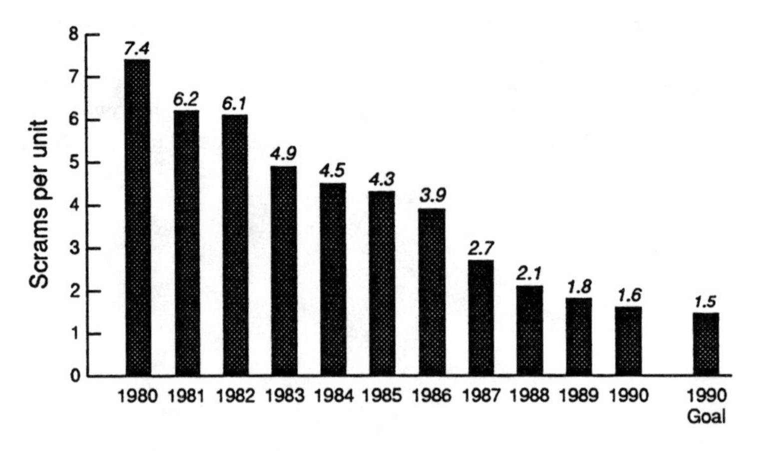

FIGURE 2-1 U.S. industry performance indicator trends through 1990 (p. 1 of 4). SOURCE: [Z. Pate, President, INPO, personal communication, 1991]

Unplanned safety system actuations

Unplanned safety system actuations include unplanned emergency core cooling system actuations and emergency AC power system actuations due to loss of power to a safeguards bus.

The industry indicator monitors the actual operation of major system components; NRC's "Safety System Actuations" indicator monitors all actuation signals whether or not the signal results in system operation.

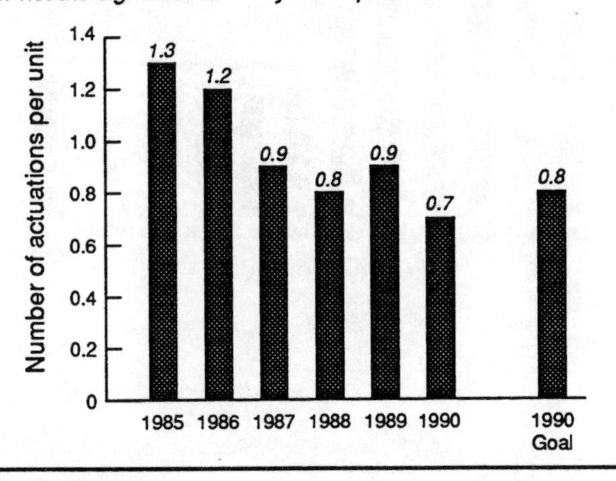

Lost-time accident rate

Lost-time accident rate is the number of worker injuries involving days away from work for every 200,000 man-hours (100 man-years) worked.

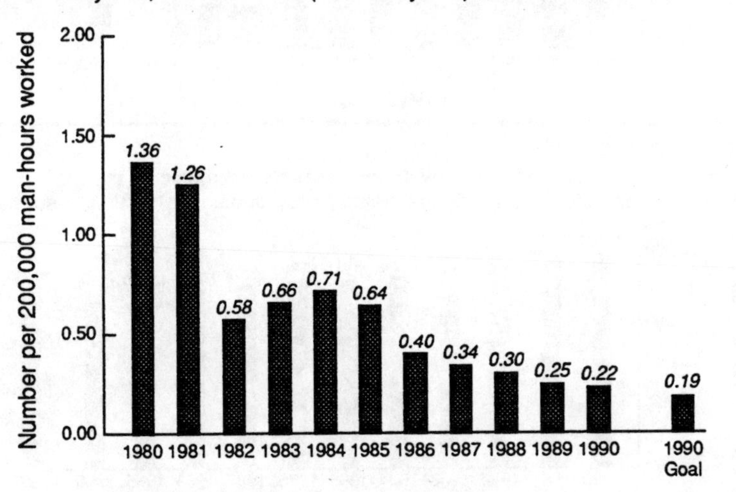

FIGURE 2-1 U.S. industry performance indicator trends through 1990 (p. 2 of 4). SOURCE: [Z. Pate, President, INPO, personal communication, 1991]

Low-level, solid radioactive waste per unit

The average annual volume of radioactive waste per unit for both boiling water and pressurized water reactors is shown on these charts.

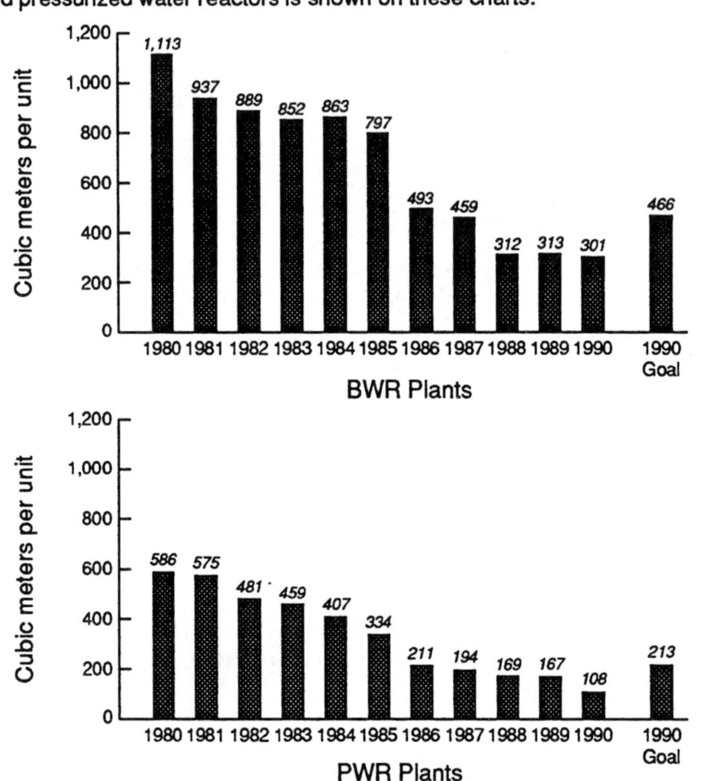

Gross heat rate

Low gross heat rate, or btu per kilowatt hour, reflects emphasis on thermal efficiency and attention to detail in the maintenance of balance-of-plant systems.

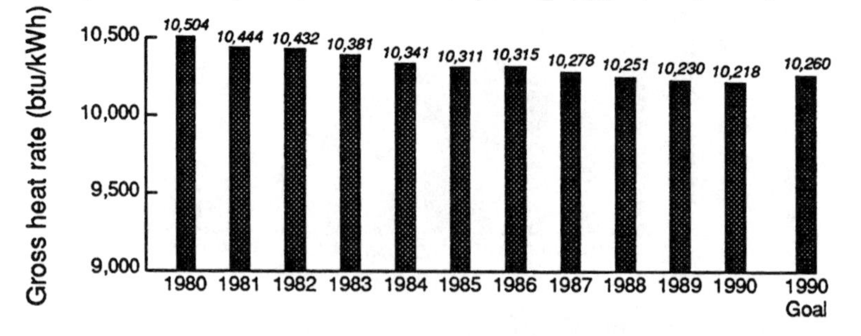

FIGURE 2-1 U.S. industry performance indicator trends through 1990 (p. 3 of 4). SOURCE: [Z. Pate, President, INPO, personal communication, 1991]

Collective radiation exposure per unit

This indicator examines the average annual collective radiation exposure in man-rem per unit for both boiling water reactors (BWRs) and pressurized water reactors (PWRs).

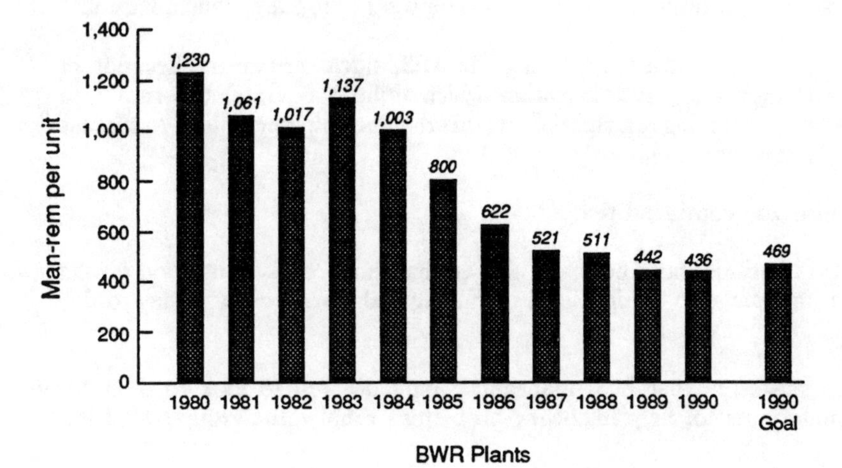

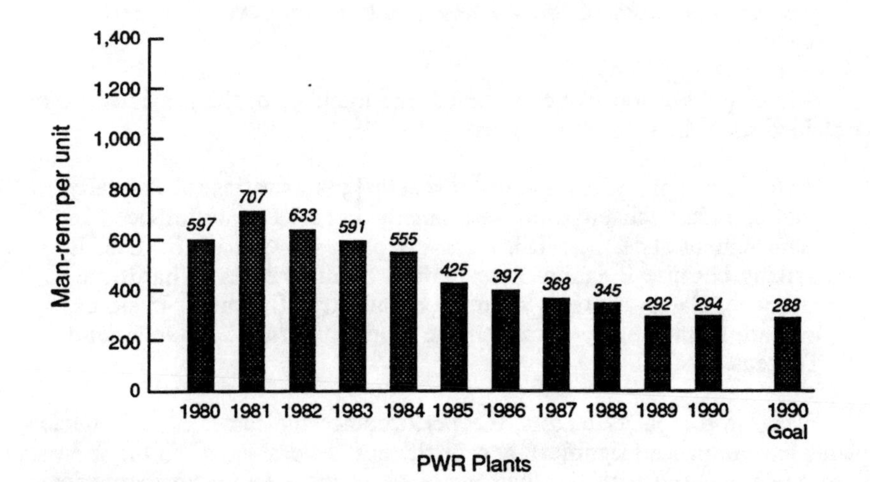

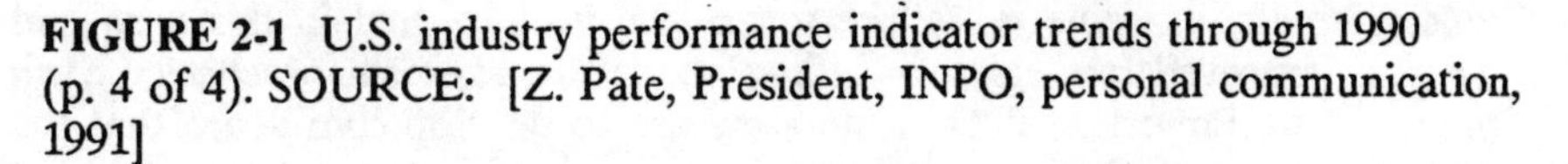

FIGURE 2-1 U.S. industry performance indicator trends through 1990 (p. 4 of 4). SOURCE: [Z. Pate, President, INPO, personal communication, 1991]

The obvious question is, "What conclusions can be drawn from a review of these and similar performance indicators?" After studying nuclear plant performance in several countries and observing the great variation in performance among U.S. plants, one group of researchers from the Massachusetts Institute of Technology [Hansen et al., 1989] concluded that

> . . . the key to improving the U.S. nuclear industry lies not in changing the system within which utilities operate, but rather in implementing managerial reforms that have proven crucial to success elsewhere.

It also was convinced that

> . . . utilities both here and abroad that show consistently good results operate with a high level of managerial involvement in day-to-day problems, . . .

and suggested that U.S. managers would do well to look to their foreign counterparts for help in solving problems. Finally, the group stated that

> . . . managers in the United States must take vigorous steps to pressure operators of the weakest plants to improve their performance.

A later publication (which involved one member of the previous group) contained the following observation:

> Performance of LWR [light water reactor] plants varies substantially among industrialized countries, largely because of differences in management style. . . . The relatively poor showing of the U.S. is striking because it cannot be explained by differences in hardware, safety regulatory systems or nuclear industry structure. It stems, instead, from the way in which the plants are run. . . .[Golay and Todreas, 1990]

Except for capacity factors, the performance indicators of U.S. nuclear plants have improved significantly over the past several years. If the industry is to achieve parity with the load factor performance in other countries, it must carefully examine its failure to achieve its own goal in this area and develop improved strategies, including better management practices. Such practices are important if the generators are to develop confidence that the new generation of plants can achieve the higher load factors estimated by the vendors.

Construction schedules and costs have been identified as serious problems for the current generation of plants. The financial community and the generators must both be satisfied that significant improvements can be achieved before new plants can be ordered. The industry itself has recognized this, and NPOC has recently issued a "Strategic Plan for Building New Nuclear Power Plants," November 1990, which discusses these and other issues.

PUBLIC ATTITUDES

Influencing Factors

It is widely accepted that public attitudes have been a growing problem for nuclear power.

> Public support for building nuclear power plants declined from strong majority support prior to TMI, to a break-even level by 1982, to four to one opposition in 1987. Support for constructing a plant in one's local area began to decline even before TMI.[Nealy, 1990]

Several factors seem to influence public attitudes:

- For the past decade or more, electricity supplies have been ample, and the public feels no sense of urgency about supporting the addition of new generating capacity of any kind.
- The public recognizes that there are alternatives to nuclear power plants to produce electric power and believes that nuclear power is more costly than many of these alternatives.
- Well publicized problems with U.S. nuclear power plants undermine the public's perception of their safety.
- The public does not have a high degree of trust in either the governmental or industrial proponents of nuclear power.
- The public has concerns about the health effects of low-level radiation.[DeBoer, 1988]
- The public is concerned that there is no safe way to dispose of high-level radioactive waste.

• Many skeptics of nuclear power (including what Alvin Weinberg has called the "articulate elites"[31][Weinberg, 1989]) believe the potential for nuclear weapons proliferation is a major threat posed by the use of nuclear power.[32]

Ambiguity of Polls

In response to the question, "On the question of nuclear power, in general, do you favor or oppose the building of more nuclear power plants in the United States?" opinion shifted between 1975 and 1988, from about 63 percent in favor, 19 percent opposed, to about 61 percent opposed, 30 percent in favor.[Harris, 1989] In a recent poll, about 75 percent of respondents said that nuclear energy was the most dangerous way to generate electricity. [Cambridge Energy Research Associates, Inc., 1990]

On the other hand, when asked "how important a role should nuclear energy play in the national energy strategy for the future?" 81 percent of those surveyed said "very important" or "somewhat important," although when asked, "If a new power plant is needed in your area, would you favor, oppose, or reserve judgment for a nuclear plant?" 59 percent said that they would reserve judgment and 23 percent that they would oppose a nuclear plant.[DOE, 1990b]

Polls, in fact, have consistently shown higher levels of opposition to building nuclear plants near respondents' own communities as compared to nuclear power development in general. Also, some disagree that the accident at TMI was a watershed event that destroyed public confidence in nuclear power. ("When public opinion is viewed over a 15-year period beginning in the early 1970s, TMI looks like little more than a small blip, which slightly accelerated a secular trend against nuclear power.")[Ganson and Modigliani, 1989] Finally, at least on nuclear power issues, interpreting the meaning of public opinion polls is difficult because widely different views seem to be supported by different polls. This result may occur because responses appear

---

[31] For example, Carl Sagan stated in the widely read Parade Magazine that " . . . there's one other problem: All nuclear power plants use or generate uranium and plutonium that can be employed to manufacture nuclear weapons."[Sagan, 1990]

[32] Non-skeptics have also commented on the risk of proliferation. For example, one senior utility executive said "We who believe that nuclear power can and should play a role in meeting mankind's future needs for energy must do everything we can to strengthen the barriers between nuclear power and nuclear weapons."[Willrich, 1985]

especially sensitive to small differences in question wording and context in the interview.

Improving Public Attitudes

In the Committee's judgment, the following developments and conditions would improve public attitudes toward nuclear power, leading to greater public acceptance.

- A recognized need for a greater electricity supply that can best be met by new large-scale baseload generating stations;[33]
- A national environmental policy leading to sanctions to reduce emissions resulting from the use of coal, oil, and natural gas in generating electricity (Of course, such a policy would also make alternative sources of electricity more attractive, including energy-efficiency improvements and renewable energy technologies.);
- Maintaining the safe operation of existing nuclear power plants, and communicating this fact in a coherent manner to the public;
- Providing the opportunity for meaningful public participation in nuclear power issues, including generation planning, siting, and oversight;[34]
- Communicating to the public in a coherent and comprehensive way the whole issue of natural and man-made low level radiation as well as that of perceived and estimated risks;
- A resolution of the high-level radioactive waste disposal stalemate; and
- Assurances that a revival of nuclear power would not materially affect the likelihood of nuclear weapons proliferation.

---

[33] Nuclear plants are presently considered not to be appropriate for service as peaking units for operational as well as economic reasons, although they can be used for load-following.

[34] As an illustration of the way greater public participation might improve public acceptance, it could be worthwhile to mention that (what are called) Local Information Commissions are established in France where nuclear power plants are located. Each Commission is chaired by an elected representative of the local population and consists of representatives from the media, unions, various associations, etc. These Commissions are regularly informed about the operation of the plant and can at any time summon the management to provide explanations on any incident which might occur. The mere existence of such Commissions assures the local population that no event seriously affecting the safety of the plant or the environment could happen without the public being immediately aware of it.[G.Vendryes, personal communication]

SAFETY ISSUES

Prevention of Nuclear Accidents

U.S. designers of nuclear plants have employed a "defense-in-depth" philosophy to ensure the safety of these plants. Defense-in-depth means the use of multiple safety systems and barriers to prevent serious accidents or mitigate their consequences. Measures intended to prevent accidents also protect the investments in the plant, the people who work in it, and the general public. Measures designed to mitigate the effects of an accident, once it occurs, are intended primarily to protect the public's health and safety.

The preceding section on public attitudes indicated that the potential hazards of nuclear power are important to the public. Although having enormously different health consequences, the accidents at TMI and Chernobyl reinforced the public's concern about the safety of nuclear power plants. The Congress recognized the necessity of assuring that peaceful applications of atomic energy would be safe when it passed the Atomic Energy Act, which states

> . . . regulation by the United States of the production and utilization of atomic energy and of the facilities used in connection therewith is necessary in the national interest to assure the common defense and security and to protect the health and safety of the public.[U.S. Congress, 1981]

Investors in nuclear plants are interested in a high level of accident prevention measures because a serious core damage accident will turn the investors' assets into a large liability. For example, TMI, Unit 2, a new plant that cost about $1 billion to build, became a liability to General Public Utilities within a few hours. The cleanup costs are estimated to have been about $1 billion, exclusive of replacement power costs.[35] Moreover, an accident in one plant, even though it causes no outside fatalities or health effects, is likely to have profound consequences for all similar plants. The

---

[35] There is no doubt that the top executives of utilities that own nuclear plants are concerned about protecting the investment. For example, Detroit Edison's new chairman stated: "Fermi represents 35 percent of our assets, 10 percent of our capacity, 20 percent of our energy. The first thing I do every morning when I come in is look at the morning report and see what's going on at Fermi. When it runs well, the company does well." The previous chairman was also reported to have said, after learning about a serious operator error at Fermi in 1985, ". . . we all found out about this very serious safety shortcoming. I was literally sick to my stomach."[Myers, 1990]

TMI accident led to an immediate, though temporary, shutdown of all similar reactors in the United States.

Public policy makers also should be concerned about the level of accident prevention measures because another accident like that at TMI in the near future would seriously affect the future of nuclear power in the United States.

While the Committee is mindful of the importance of the mitigation of accident consequences to protect the public, the safety discussion in this section has focused on the prevention of accidents. In the next section, the safety policy of NRC as it relates to the protection of the public is discussed.

Nuclear Regulatory Commission's Safety Policy

The Atomic Energy Act, as amended, authorizes and mandates NRC to regulate commercial use of nuclear energy to protect the public health and safety. NRC must therefore base its decisions on public health and safety considerations rather than on the economic impact on a utility or the industry.

NRC published the current safety policy, "Safety Goals for the Operations of Nuclear Power Plants," in November 1988.[10 CFR, Part 50, 1988] This policy does not contain required numerical values for core melt frequency, emergency core cooling system failure rates, or other performance characteristics. Instead, it sets forth qualitative safety goals and high level quantitative objectives to protect the public. (See Note with 10 CFR Part 50 at the end of this chapter.) However, the Commission has recently stated

> A core damage probability of less than 1 in 10,000 per year of reactor operation appears to be a very useful subsidiary benchmark in making judgments about that portion of our regulations which are directed toward accident prevention.[Chilk, 1990a]

With respect to this benchmark, the NRC staff has stated

> If each of the current population of approximately 100 plants had a calculated core damage frequency approximating this overall mean value [i.e., 1×10^{-4} per reactor year], it would imply the overall occurrence of such events, on average, at a frequency of about once in a hundred years, a time interval larger than the expected lifetime of any single plant.[Stello, 1989]

The design requirements for the advanced reactors are more stringent than the NRC safety goal policy.

Utility Improvement and Self-Regulation

Inherent in the Atomic Energy Act, as amended, and in the licensing process for utilities is the idea that the primary responsibility for the safety of commercial nuclear power plants rests with the operator (the licensee). The important safety role of federal regulation must not be allowed to detract from or undermine the accountability of utilities and their line management organizations for the safety of their plants.

Over the past decade, utilities have steadily strengthened their ability to meet this responsibility. Their actions include the formation and support of industry institutions, including INPO. Self-assessment and peer oversight through INPO are acknowledged to be strong and effective means of improving the performance of U.S. nuclear power plants. This U.S. utility industry organization has recently been emulated on a worldwide scale by the formation of the World Association of Nuclear Operators (WANO).

The Committee believes that such industry self-improvement, accountability, and self-regulation efforts improve the ability to retain nuclear power as an option for meeting U.S. electric energy requirements. The Committee recommends that NRC encourage such industry initiatives.

Safety of Existing Reactors

There are three basic methods that provide complementary insights into the safety of a reactor or set of reactors: operational history, inspections for compliance with applicable standards, and probabilistic risk analysis.

In over 1,400 reactor years of U.S. commercial reactor operation [NRC, 1990b], there have been one core melt accident (TMI), one serious fire that threatened core damage (Brown's Ferry), and several serious system or component failures in commercial LWRs.[36] None has led to significant off-site releases of radioactive material.

NRC administers and directs an accident sequence precursor (ASP) program. This program was established at the Nuclear Operations Analysis

---

[36] The Browns Ferry fire was discussed in the U.S. Nuclear Regulatory Commission's Annual Report for 1975, and the Accident at Three Mile Island was the subject of the report of a Presidential Commission in October 1979. A possible recent example of a serious system failure was the loss of control rod position indication, feedwater control system, plant computers, and some plant lighting as a result of a main transformer fault at the Nine Mile Point Unit 2 nuclear reactor in August, 1991.[NRC, 1991b,c]

Center at Oak Ridge National Laboratory in 1979 to provide a means of evaluating the significance of operational experience. Under this program, operational experience reported by the U.S. commercial nuclear power plants was reviewed to identify and categorize precursors to potential severe core damage accidents.[NRC, 1989b] According to NRC's report:

> The operational events selected in the ASP Program form a unique database of historical system failures, multiple losses of redundancy, and infrequent accident initiators. These events are useful in identifying significant weaknesses in design and operation, for use in analysis of industry performance, and for use in probabilistic risk assessment-related studies . . . operational occurrences that involve portions of postulated core damage sequences are identified and evaluated. Event tree models and probabilistic risk assessment techniques are used to put the reported data in perspective for evaluation. The event trees model plant equipment that could affect, or could be used to mitigate, the event being evaluated, as well as human actions. This method allows quantitative estimates of the significance of the event in terms of a conditional core damage probability.[37]
>
> The breakdown of precursors in 1984 through 1988 by conditional core damage probability is shown in [Table 2-12]. In 1985, there was one precursor with conditional core damage probability in the 1E-2 range and one precursor with this probability in the 1E-3 range. The 1985 precursor with conditional core damage probability in the range of 1E-2 was an operational event at Davis-Besse Nuclear Power Station involving a complete loss of feedwater. The precursor with conditional core damage probability in the 1E-3 range was an operational event at Edwin I. Hatch Nuclear Plant, Unit 1, involving a stuck-open safety relief valve and a loss of the high-pressure coolant injection system and the reactor isolation cooling system.
>
> In 1986, there were two precursors with conditional core damage probability in the 1E-3 range. The precursors were operational events at Catawba Nuclear Station, Unit 1 . . . , and at Turkey Point Plant, Unit 3. . . . The event at Catawba Unit 1 was a small-break loss-of-coolant accident involving a guillotine rupture of the letdown line. The event at Turkey Point Unit 3 involved a reactor trip with a stuck-open pressurizer relief valve. There were no precursors with

---

[37] The definition of "conditional core damage probability" provided by the NRC staff follows: "[It] is the likelihood that an event or condition will result in core damage given actual observed initiating conditions and degradation and failures of equipment needed to mitigate the event."[Jordan, 1991]

TABLE 2-12 Number of Precursors and Associated Conditional Core Damage Probabilities[a] (from NRC's Accident Sequence Precursor Program)

Conditional Probability[b]	Number of Precursors by Year 1984	1985	1986	1987	1988
1E-2	0	1	0	0	0
1E-3	1	1	2	0	0
1E-4	15	8	4	10	7
1E-5	8	13	7	9	14
1E-6	8	16	5	14	11

[a] The definition of "conditional core damage probability" provided by the NRC staff follows: "[It] is the likelihood that an event or condition will result in core damage given actual observed initiating conditions and degradation and failures of equipment needed to mitigate the event." [Jordan, 1991]

[b] 1E-2 means 10^{-2}, or 0.01, etc.

SOURCE: [NRC, 1989d]

conditional core damage probability greater than 1E-3 in 1987 or 1988.

In 1988, four of the seven precursors with the highest conditional core damage probability (i.e., 1988 precursors having an estimated conditional core damage probability greater than 1E-4) involved common mode failures; another event involved potential common mode failures. These data illustrate the importance of common mode failures to reactor safety and the need for continued vigilance in the areas of maintenance, inspection, and testing of safety equipment.[NRC, 1989b]

The data in Table 2-12 indicate a slightly declining frequency of occurrence of precursors with relatively high conditional probabilities, suggesting that safety may be improving.

Inspection is used by NRC to evaluate a plant in relation to the large body of NRC regulations and license commitments. If a plant is found by inspectors to meet the regulations, it is deemed "safe." Few plants have been ordered shut down because NRC found them unsafe, although examples include Pilgrim and Peach Bottom. NRC has never permanently shut down a licensed plant on safety grounds.

Probabilistic risk analysis (PRA) has been used increasingly since 1979 when NRC endorsed this technique following a congressionally mandated review of the Atomic Energy Commission's Reactor Safety Study.[38][Chilk, 1979] The review recommended greater use of PRA.[Lewis, 1978]. As a result, PRAs have been performed for many U.S. nuclear plants. By the mid-1980s, new methods for analyzing severe accidents had evolved, leading NRC to reassess the risks of such accidents in five commercial nuclear plants. The results are presented in the latest version of NUREG-1150 and numerous supporting documents.[NRC, 1989c] NUREG-1150, entitled Severe Accident Risks: An Assessment for Five U.S. Nuclear Power Plants, is a major step forward in PRA methods. The major advancements in NUREG-1150 methodology are the inclusion of an uncertainty analysis based on the use of expert opinion to develop parameters and probability distributions where there is insufficient experimental and analytical data, and the inclusion of external initiating events in two cases (Surry and Peach Bottom). The elicitation of expert opinion is a formalized process so that the assumptions and approximations employed by the risk analysts become explicit to all who read the analyses. There are two issues regarding this procedure, however; the question of just who is an expert on a given issue, and the data upon which the experts base their opinion. In NUREG-1150, these issues are important when considering how the methodology and results will be used, and in understanding the limitations of this methodology.[39]

In 1989, the NRC staff stated:

> Available PRA evidence to date suggests that current plants, on the whole, probably are configured such that the overall mean core damage frequency is probably near but still somewhat above 10^{-4} per year.[Stello, 1989]

However, NUREG-1150, in 1990, indicated that this estimate may be pessimistic. Although "NUREG-1150 is not an estimate of the risks of all

---

[38] Specifically, NRC stated: "Taking due account of the reservations expressed in the Review Group Report and in its presentation to the Commission, the Commission supports the extended use of probabilistic risk assessment in regulatory decisionmaking."[Chilk, 1979]

[39] The Advisory Committee on Reactor Safeguards has cautioned, "Since there is a dearth of information concerning many of the phenomena that determine severe accident progression, expert elicitation was used most extensively in the Level 2 portion of the PRAs.... However, with insufficient information there can be no experts. Thus, use of the term 'expert opinion' in a description of some of the Level 2 work may be misleading."[NRC, 1990g]

commercial nuclear power plants in the United States," it did provide "a snapshot in time of severe accident risks in five specific commercial nuclear power plants."

So-called external events (e.g., seismic and fire) were analyzed for only two of the five plants (Surry and Peach Bottom). Two widely divergent predictions for the seismic hazard curve existed, one prepared by Lawrence Livermore National Laboratory by consulting a large group of experts [Bernreuter et al., 1989], the other by the EPRI [Seismicity Owners Group and EPRI, 1986] employing somewhat different methods of using expert opinion. NRC chose to report the core melt frequencies associated with each seismic hazard curve independently, rather than average them in some fashion.

The mean core melt frequencies reported in NUREG-1150 for these five commercial nuclear power plants are reproduced in Table 2-13.

TABLE 2-13 Mean Core Melt Frequency (Reactor Year^{-1})

	INTERNAL EVENTS	EXTERNAL EVENTS	
		LLNL Seismic	EPRI Seismic
Surry	4.0E-5	1.3E-4	3.6E-5
Peach Bottom	4.5E-6	9.7E-5	2.3E-5
Sequoyah	5.7E-5		
Grand Gulf	4.0E-6		
Zion	3.4E-4*		

*Recent changes in equipment and procedures now lead to a predicted mean core melt frequency of 6E-5 for the Zion Plant, according to NUREG-1150

SOURCE: [NRC, 1989c]

It is noted that all commercial nuclear power plants that have not had a prior PRA performed are required to undertake a PRA, or an equivalent systematic evaluation of the risk of core melt and of significant radioactivity release, in order to identify any plant specific "outliers" that might be making too large a contribution to risk.

One of the key utility design requirements for advanced LWRs (discussed later in Chapter 3 of this report) is for the core melt frequency to be less than 1 in 100,000 years of operation as estimated by PRA.[40]

PRA has proven to be of greatest value for comparison and insight. NRC used PRA in the early 1980s to review the safety of the Indian Point reactors by comparing them with other reactors. PRA also has provided insights into previously unseen problems. As NRC Chairman Carr recently observed: "... virtually every probabilistic risk assessment (PRA) performed has led to some modifications in plant design or operational practices that would reduce the estimated severe core damage frequency."[Carr, 1990] However, both NRC and its Advisory Committee on Reactor Safeguards have expressed concern that, in view of the large uncertainties in PRA, the results not be misused. NRC Chairman Carr also stated that

> . . . simple estimates are subject to much uncertainty inherent in projecting core damage probabilities; these averages are driven by plants that may have much higher core damage frequency than the majority and for this and other reasons, are subject to potential misuse.[Carr, 1990]

With respect to NUREG-1150, NRC's Advisory Committee on Reactor Safeguards (ACRS) recommended that

> . . . its results should be used only by those who have a thorough understanding of its limitations.[NRC, 1990g]

Earlier, in commenting on approaches to implement NRC's safety goal policy, the ACRS observed

---

[40] The discussion in this and in the previous section on safety goals is of core melt accidents. The containment building is expected to provide significant additional protection against radiation release for most accident scenarios. Consistent with this concept, the NRC, in addressing goals for evolutionary LWRs, "approved the overall mean frequency of a large release of radioactive material to the environment from a reactor accident as less than one in one million [$1x10^{-6}$] per year of reactor operation. The Commission has not agreed on a definition of a large release. . . . "[Chilk, 1990b]

> . . . it is universally agreed that the 'bottom line' estimates . . . are among the weakest results of a PRA.[NRC, 1988a]

and

> We do not believe that probabilistic risk assessment (PRA) . . . is sufficiently developed to be used to make narrowly differentiated decisions about specific plants. . . . the search for risk outliers for individual plants should be performed. We believe that detailed qualitative information on plant characteristics and behavior is an important result of such a search, but that quantitative information (such as core melt frequency estimates for an individual plant) developed by a PRA is less robust.[NRC, 1987a]

and

> We note that there must be recognition of important limitations in the implementation of the Safety Goal Policy. These limitations are essentially those of the PRA methodology used, and are caused by a fundamental inability to accurately predict and calculate precise values of risk. Variability in data, uncertainty about applicability of data, imperfect understanding of important physical phenomena, and inevitable incompleteness in analysis all contribute to this limitation. [NRC, 1987a]

While PRA is not a perfect tool for assessing risk, it provides valuable methods and is currently used by vendors and some utilities to evaluate design modifications. NRC has also requested that an integrated plant examination be performed at all U.S. nuclear plants using PRA techniques. The purposes of this examination are:

> . . . for each utility (1) to develop an appreciation of severe accident behavior, (2) to understand the most likely severe accident sequences that could occur at its plant, (3) to gain a more quantitative understanding of the overall probabilities of core damage and fission product releases, and (4) if necessary, to reduce the overall probabilities of core damage and fission product releases by modifying, where appropriate, hardware and procedures that would help prevent or mitigate severe accidents. It is expected that the achievement of these goals will help verify that at U.S. nuclear power plants severe core damage and large radioactive release probabilities are consistent with the Commission's Safety Goal Policy Statement.[NRC, 1988b]

After examining the available information, the Committee has reached the following conclusions.

- The risk to the health of the public from the operation of current reactors in the United States is very small. In this fundamental sense, current reactors are safe.
- A significant segment of the public has a different perception, and also believes that the level of safety can and should be increased.
- As a result of operating experience, improved operator and maintenance training programs, safety research, better inspections, and productive use of PRA, safety is continually improved. In many cases these improvements are closely linked to improvements in simplicity, reliability, and economy.

Industry plans for advanced reactors include safety requirements that exceed those of current plants.[EPRI, 1986]

HIGH-LEVEL RADIOACTIVE WASTE DISPOSAL

Lack of resolution of the high-level waste problem jeopardizes future nuclear power development. First, there are enough arguments made against nuclear power based on the lack of resolution of the high-level waste disposal issue that this constitutes a major cause of the public's unfavorable perception of nuclear power.[41] Second, state regulation may prohibit further nuclear power development until the high-level waste disposal issue is resolved. For example, California law prohibits the construction of more nuclear plants in that state until the California Energy Commission certifies that the high-level waste problem is solved.

In the United States, DOE has been assigned the task of siting, constructing, and eventually operating a geologic high-level radioactive waste repository. The work is being funded by ratepayers through a special surcharge on electricity generated at nuclear power plants. DOE now estimates that this geologic waste repository will not be ready to receive spent reactor fuel before about 2010. Even this date is in doubt, given the legal and regulatory problems and the political and technical uncertainties that have arisen regarding the identified Yucca Mountain site.[National Research Council, 1990] These problems are exacerbated by the requirement that, before operation of a repository begins, DOE must demonstrate to NRC that the

---

[41] For example, a paper discussing the case against reviving nuclear power states "The daunting problems of nuclear waste disposal and nuclear materials proliferation grow ever more indomitable as governments fail to come up with solutions and the materials themselves accumulate."[Flavin, 1988]

repository will perform to standards established by the U.S. Environmental Protection Agency (EPA), which limit the release of radionuclides to specific levels for 10,000 years after disposal. [EPA, 40 CFR 191] NRC's staff has strongly questioned the workability of these quantitative requirements, as have the National Research Council's Radioactive Waste Management Board and others.[42] For example

> The . . . [National Research Council Board on Radioactive Waste Management] believes that this use of geological information and analytical tools--to pretend to be able to make very accurate predictions of long-term site behavior--is scientifically unsound.

The Board also wrote:

> The United States appears to be the only country to have taken the approach of writing detailed regulations before all of the data are in. As a result, the U.S. program is bound by requirements that may be impossible to meet.

and

> . . . the demand for accountability in our political system has fostered a tendency to promise a degree of certainty that cannot be realized. [National Research Council, 1990]

EPA's criteria also were criticized by the Nuclear Waste Technical Review Board set up by the Congress to review the high-level waste program:

> . . . the release limits [in the draft revision of 40 CFR 191, the EPA high-level waste regulation] appear very conservative and inconsistent with present day regulatory practice and scientific consensus. [Nuclear Waste Technical Review Board, 1990]

In addition, the criteria have been criticized by EPA's own Scientific Advisory Board.[Collier, 1984]

The Committee concludes that the EPA standard for disposal of high-level waste will have to be reevaluated to ensure that a standard that is both adequate and feasible is applied to the geologic waste repository.

---

[42] 10 CFR 60, promulgated by NRC, might also present difficulties, depending in part on how NRC's staff seeks assurance that the EPA standards and NRC's own requirements have been met, particularly for events such as intrusion and climate changes. The Committee did not analyze the implications of 10 CFR 60.

In the meantime, storage pools at some operating reactors are nearing their licensed capacities. However, dry storage of spent fuel could alleviate this immediate problem at most reactors, and such storage has been judged to be adequate for many decades.[NRC, 1990c] The Atlantic Council of the United States has recommended that the electric utility industry "develop plans for intermediate storage at plant sites or special sites in order to assure continued operation of their nuclear plants if the DOE deadlines are not met." [Atlantic Council of the United States, 1990]

The Committee believes that the legal status of the Yucca Mountain site for a geologic repository should be resolved soon, and that DOE's program to investigate this site should be continued. In addition, a contingency plan must be developed to store high-level radioactive waste in surface storage facilities pending the availability of the geologic repository. However, by current law the federal government cannot construct a temporary above-ground storage facility (Monitored Retrievable Storage, or MRS) until the Commission has issued a license for the construction of a repository.[43]

LICENSING AND REGULATION

The New Licensing Rule

An obstacle to continued nuclear power development in the United States has been the uncertainties in NRC's licensing process. The new licensing rule, 10 CFR Part 52, was intended to improve this process. The rule provided for (1) certification of reactor designs, (2) early NRC approval of nuclear power plant sites, and (3) a combined construction and operating license for applications for certified reactors on pre-approved sites.[GSA, 1990] The rule was designed to deal with practically all licensing issues in the initial stages of the project, leaving to the end only relatively narrow issues such as whether the plant had been built in accordance with the license. The Commission believed that the new rule went as far as its legislative authority

---

[43] Section 148 (d)(1) of the Nuclear Waste Policy Act of 1982, amended in 1987, provides that "construction of such [a monitored retrievable storage] facility may not begin until the Commission has issued a license for the construction of a repository. . . ." However, DOE's National Energy Strategy indicates that Congress will be requested to enact legislation to address, among other things, " . . . the siting and operation of the MRS [monitored retrievable storage] facility, which is needed to begin Federal acceptance of spent nuclear fuel by 1998. Progress on the siting and licensing of the MRS facility should be independent of the schedule for siting and licensing the repository."[DOE, 1991]

permitted in establishing a "one step" licensing regime. However, the new rule was challenged in court. The basis of the challenge was that, by increasing the number of issues decided early in the process so as to largely eliminate the possibility of a second hearing after construction, the Commission had violated the Atomic Energy Act. The United States Court of Appeals for the District of Columbia Circuit ruled, on November 2, 1990, that

> . . . the plain language of Section 185 [of the Atomic Energy Act] requires the Commission to make a post-construction, pre-operation finding that a nuclear plant will operate in conformity with the Act and that the plain language of Section 189(a) requires the Commission to provide an opportunity for a hearing to consider significant new information that comes to light after initial licensing and that implicates the Commission's finding obligations under Section 185. Accordingly, we find that two subsections of the regulations are inconsistent with the statute. We thus vacate 10 C.F.R. Section 52.103(b) and 10 C.F.R. Section 52.103(c); we uphold the remainder of the regulations against petitioners' various challenges.

The Court concluded that the Commission's

> . . . 'rulemaking power is limited to adopting regulations to carry into effect the will of Congress as expressed in the statute.' . . . Thus, the ultimate responsibility for such reforms as embodied in Sections 52.103(b) and (c) lies not with the Commission, but with the Congress.[U.S. Court of Appeals for the District of Columbia, 1990]

On March 27, 1991 the U.S. Court of Appeals for the District of Columbia vacated this November 2nd decision and chose to address, *en banc*, these and other licensing issues.[Energy Daily, 1991]

It is likely that, if the possibility of a second hearing is to be reduced or eliminated, legislation will be necessary. The nuclear industry is convinced that such legislation will be required to increase utility and investor confidence to retain nuclear power as an option for meeting U.S. electric energy requirements.[NPOC, 1990] The Committee concurs.

There are important questions about the level of detail required to certify a new reactor design. For example, one portion of 10 CFR Part 52 that was discussed earlier and was not overturned by the court says that the design of an evolutionary LWR proposed for certification should be "essentially complete." The meaning of this term has been clarified by a recent NRC policy statement.[Chilk, 1991]

The Committee views a high degree of standardization as very important for the retention of nuclear power as an option for meeting U.S. electric energy requirements. Such an approach has been shown to be effective in France, in the Konvoi plants in Germany, and in Canada. The long-term success of standardization will depend on a determination by new owners to insist on standardized designs, and their willingness to maintain a high degree of standardization during construction and throughout the life of the plant.

The Nuclear Regulatory Commission

Impact of Advanced Reactors

Earlier in this chapter the Committee stated that NRC's regulations should be critically reviewed and modified (or replaced with a more coherent body of regulations) for advanced reactors. In addition, some of the advanced reactor technologies discussed later in this report will make new demands on NRC. For example, the licensing of an advanced liquid metal reactor and *in situ* reprocessing may raise new licensing issues and may require reopening of the GESMO (Generic Environmental Statement on Mixed Oxide) proceeding. In addition, NRC may have to address the licensing of new institutional arrangements because of reprocessing, concerns about diversion of sensitive nuclear materials, and lack of utility experience with the technology.

Relations with Licensees

Nuclear plant licensees have been critical of the Commission as evidenced by comments received in 10 major areas in a recent survey of utilities:[44] [NRC, 1990e]

---

[44] The following statement, which was included in NUREG 1395 [NRC, 1990e], provides perspective for this survey:

> In reading the summary and the specific licensee comments presented in this survey, it must be borne in mind that these views are not intended as a balanced portrayal of the impact of NRC activities. The staff sought out licensee observations of problems and the perceptions of problems in NRC's activities rather than comments on the benefits or the positive effects of agency regulatory activities. It is not surprising, therefore, that this survey portrays a one-sided view of NRC activities. In some cases, the perceptions and opinions given are at variance with the staff's understanding of the facts. Nonetheless, the report presents the unvarnished views of the wide range of licensee representatives who talked with the staff.

- Licensees believe that the Commission issues so many new requirements that it is attempting to manage the utilities' resources rather than to regulate the industry.
- The Commission was accused of untimely reviews of licensee submittals relating to technical issues, of issuing technical specifications of low quality, and of providing inadequate provisions for appeal on technical matters.
- Objections were raised that Commission inspectors impose many backfits, unauthorized by the Commission, by setting successively higher standards of performance.
- Licensees object to NRC's Systematic Assessment of Licensee Performance (SALP) process, arguing that (a) it is an improper mechanism for obtaining improved performance, and (b) public and outside organizations misuse and misinterpret SALP results. They argue it is too subjective and not uniformly applied among the regions.
- Objections were raised about the collective impact of oversight by multiple organizations such as the Commission, state safety inspectors, insurance inspectors, and personnel from INPO.
- Although all utilities supported recent changes to the operator requalification examination process, they were concerned that operators are not permitted to function in the simulator examination process as they normally do on shift, that examiner standards change continually, and that too many organizations are involved in requalifications.
- Licensees complained that the Commission takes enforcement action for violations and new generic requirements for which corrective action has been taken or is planned. They expressed fears that challenges by the utilities to such actions would result in lower ratings in their performance assessments.
- Complaints were made that the Commission's thresholds for reporting significant events are too low, that conflicts exist in the documents governing reporting requirements, and that reporting may impair licensees' ability to respond to an event.
- Licensees expressed reluctance to raise issues about Commission actions for fear of retaliation.
- Issues were raised about the qualifications and training of commission personnel.

NRC's staff is reviewing such complaints to see what can be done to improve and reduce requirements, to apply Commission rules consistently, and to improve performance of Commission personnel. The Commission's survey of NRC employees, published in mid-1990, revealed an underlying general observation that licensees are extremely sensitive to NRC activities and sometimes acquiesce to avoid confrontations. In addition, three other themes were:

- NRC neither adequately considers cumulative impacts on licensees of requirements it issues nor identifies priorities for such requirements;
- NRC significantly impacts licensees by the volume and scheduling of its on-site activities; and
- NRC's continued loss of experienced professionals has depleted its knowledge base and, in some instances, unnecessarily impacted licensees.

Changes to the regulatory program are being considered by NRC as a response to the above findings.[NRC, 1990d]

The Committee concludes that NRC should improve the quality of its regulation of existing and future nuclear power plants, including tighter management controls over all of its interactions with licensees and consistency of regional activities. Industry has proposed such to NRC.[Lee, 1991]

The Committee encourages efforts on both sides to reduce reliance on the adversarial approach to issue resolution.

Possible Conflicts of Interest

NRC's staff is often required to investigate its own role when serious incidents at nuclear plants occur, which some believe represents a serious conflict of interest.[Lewis, 1986] Furthermore, experience shows that the staff managed investigations often do not identify NRC actions or inactions as among the root causes of incidents.[Lewis, 1986]

One approach that has been proposed to correct the above problems is the formation of an accident investigation board separate from the NRC staff. Such a group would be modelled generally after the National Transportation Safety Board (NTSB).[Union of Concerned Scientists, 1987; Lewis, 1986] NTSB is independent of the Federal Aviation Administration (FAA), so it can criticize FAA's role in an accident, such as a controller error, as well as the aviation industry's role. Such a Board could help assure objective illumination of the role played by NRC's personnel and processes in nuclear accidents or near-misses. It could also help assure that recommendations for corrective measures address NRC actions or inactions.

There is considerable Congressional testimony over the last decade in favor of an NTSB-like organization to examine nuclear power accidents. Rep. Udall's Interior Committee, for example, has held hearings on this concept.

Professor H. Lewis, ACRS member, has testified many times in support of this concept. For example, in 1986 he told the House Interior Committee's Subcommittee on Energy and the Environment that

> . . . the best way to learn how to distinguish the real precursors to major accident sequences from the distractions is to learn systematically from operating experience, and that that requires an investigation that is disjoint from the issues of regulatory self-interest or of imposition of the necessary fixes. In both of these matters, the regulatory agency, whichever it is, is not above suspicion. (Indeed, the prototype for such boards, the National Transportation Safety Board, was once part of the regulatory agency, but experience demonstrated the prudence of separation, as it will here, in time.) [Lewis, 1986]

NTSB always includes industry experts on the investigative teams, thereby getting the most knowledgeable people.

The Committee believes that the NTSB approach, as a substitute for the present NRC approach, has merit. In view of the infrequent nature of the activities of such a committee, it may be feasible for it to be established on an ad hoc basis and report directly to NRC's chairman. Before the establishment of such an activity, its charter should be carefully defined, along with a clear delineation of the classes of accidents it would investigate. Its location in the government and its reporting channels should also be specified.

State Regulation of Nuclear Power[45]

The Committee believes that the trends in government involvement in regulation are to transfer authority from the federal to state and local governments. The Committee has not explicitly addressed the long-term implications for nuclear power of these changes, except for the changing role of states in safety and economic regulation.[46]

---

[45] The Committee notes that state regulatory authorities have limited influence over federal power marketing administrations or municipal utilities.

[46] A general reference for state-federal interactions that involve nuclear regulation was prepared in 1987 under the auspices of Lawrence Livermore National Laboratory.[Pasternak and Budnitz, 1987]

State Safety Regulation

Although the Atomic Energy Act of 1954 assigned the role of safety regulation to the Atomic Energy Commission and its successor since 1975, NRC, several states have established substantial programs for safety oversight of nuclear power plants. Some utilities are concerned that aggressive state oversight programs could complicate an already difficult management job, reduce efficiency, increase costs, and perhaps adversely affect safety.[Inside NRC, 1990] However, the states with current programs have not attempted to take action in areas reserved for federal authorities, and, in general, state personnel coordinate their activities with local NRC personnel. The Committee sees the possibility that existing state programs might expand and that additional states may engage in safety oversight activities.

State Economic Regulation

The states have primary authority for the economic regulation of the production and retail sale of electrical power within their borders. Among the most important decisions of state public utility commissions are those relating to what capital expenditures may be incorporated in the rate base and recovered from customers with a return on investment.

Some nuclear power proponents contend that, since utilities have a relatively low allowed rate of return, they must have a high level of assurance of full cost recovery. However, any agreements made in advance are unlikely to incorporate guarantees of recovery of costs that substantially exceed costs for alternative ways to provide the same service to ratepayers. Thus, unless the problems that have led to the current high construction costs and cost overruns of nuclear plants are solved, limited assurances are not likely to be of much value.

One remedial response would be enactment by the states of the Utility Construction Review Act offered by the Council of State Governments' Committee on Suggested State Legislation in July 1990. This legislation would facilitate the construction of electricity generating power plants that state regulators have authorized as necessary. It would permit periodic approvals of completed construction work on utility facilities and assured rate recovery (absent fraud, concealment, or gross mismanagement) for approved expenditures. Similar proposals by others have been called rolling prudency reviews. The concept of state-utility shared responsibility would also apply to a continuing evaluation of the need for power, so that if circumstances changed, the state public service (or utility) commissions would be obligated to immediately notify the utility building a new plant which may no longer be needed. In this case, the legislation would again permit recovery, through rates, of a utility's investment in the delayed or cancelled facility up to the time of notification by the public service (or utility) commission.[Committee on Suggested Legislation, 1991] The Committee believes that enactment of such legislation could remove much of the investor risk and uncertainty currently associated with state regulatory treatment of new power plant construction, and could therefore help retain nuclear power as an option for meeting U.S. electric energy requirements.

On balance, however, unless many states adopt this or similar legislation, it is the Committee's view that substantial assurances probably cannot be given, especially in advance of plant construction, that all costs incurred in building nuclear plants will be allowed into rate bases. The solution to the problem of recovering construction costs must begin with the nuclear industry. The Committee believes that greater confidence in the control of costs can be realized with plant designs that are more nearly complete before construction begins, plants that are easier to construct, use of better construction and management methods, and business arrangements among the participants that provide stronger incentives for cost-effective, timely completion of projects.

Some state public utility commissions have placed the nuclear plants they regulate under incentive systems to reward utilities for plant performance above specified levels and to penalize them for plant performance below these levels. In early 1990, a total of 73 nuclear plants in 18 states were operating under performance incentive systems that use such indicators as equivalent availability factor, fuel costs (or replacement power costs), and construction costs.[Inside NRC, 1990] In Massachusetts, the Boston Edison Company's Pilgrim plant operates under incentives primarily based on capacity factor, but also on NRC's SALP process. In addition, there are much smaller incentives related to narrowly focussed performance indicators such as number of scrams and number of safety system actuations.[Boston Edison Company, 1990] The economic effects of these provisions are uncertain, but, according to one report, the Massachusetts Attorney General's office estimated that Boston Edison's maximum annual revenue increase and loss under them would be $4.5 million and $19 million, respectively.[Inside NRC, 1989]

Industry representatives have argued that incentive arrangements using SALP or performance indicators (other than long-term capacity factor) have the potential to compromise safety.[Inside NRC, 1990] In a policy statement NRC has expressed concern about the states' use of the SALP system and other indicators for economic incentives. The policy statement recognized that state regulatory actions can have either a positive or negative impact on public health and safety, and specifically identified the approaches that are of particular concern (e.g., inappropriate reliance on SALP scores). NRC's policy is to continue monitoring incentive programs consistent with the belief that they should not create incentives to operate a plant when it should be shut down for safety reasons.[NRC, 1991d]

It is the Committee's expectation that state incentive programs will continue for nuclear power plant operators. Properly formulated and administered, these programs should improve the economic performance of nuclear plants, and they may also enhance safety. However, they do have the potential to provide incentives counter to safety. The Committee believes that such programs should focus on economic incentives and avoid incentives that can directly affect plant safety.

REFERENCES

Ahearne, J. F. 1989. Will Nuclear Power Recover in a Greenhouse? Discussion Paper. Resources for the Future. ENR89-06. Washington, D.C. May.

Ahearne, J. F. 1988. A Comparison Between Regulation of Nuclear Power in Canada and the United States. Prog. in Nuc. Energy. 22:215-229.

Allison, G., and A. Carnesale. 1983. The Utility Director's Dilemma: The Governance of Nuclear Power. Uncertain Power, The Struggle For A National Energy Policy. Pergamon Press Inc. New York. 134-153.

Atlantic Council of the United States. 1990. Energy Imperatives for the 1990s. Report of the Atlantic Council's Energy Working Group. Policy Paper. March.

Bacher, P., and M. Chapron. 1989. Nuclear Units Under Construction. Revere Générale Nucléaire, International Edition, May/June.

Bernreuter, D. L., et al. 1989. Seismic Hazard Characterization of 69 Nuclear Power Sites East of the Rocky Mountains. Lawrence Livermore National Laboratory. NUREG/CR-5250. January.

Birkhofer, A. 1991. (Lehrstuhl für Reaktordynamik und Reaktorsicherheit), Technische Universität München, Federal Republic of Germany. Provided to the National Research Council's Committee on Future Nuclear Power Development in February.

Blair, P. (Office of Technology Assessment, U.S. Congress.) 1990. Reflections on U.S. Electricity Demand and Capacity Needs. The Aspen Institute Policy Issue Forum, The Electricity Outlook. Aspen, Colorado. July 18-22, 1990.

Boston Edison Company. 1990. Viewgraph: Getting A Competitive Edge, DPU Settlement Performance Incentives. Received by Archie L. Wood.

Cambridge Energy Research Associates, Inc. 1990. Energy and the Environment: The New Landscape of Public Opinion.

Cantor, R., and J. Hewlett. 1988. The Economics of Nuclear Power, Further Evidence on Learning, Economies of Scale, and Regulatory Effects. Resources and Energy 10.

Carr, K. M., Chairman, NRC. 1990. Letter to Albert L. Babb, University of Washington, June 14, 1990.

Cavanagh, R. 1986. Least Cost Planning Imperatives for Electric Utilities and Their Regulators. Harvard Environmental Law Review 299. 10:2.

Chilk, S. J., Secretary of the Commission. 1991. Memorandum for James M. Taylor, Executive Director for Operations. Subject: SECY-90-377-Requirements for Design Certification Under 10 CFR Part 52. February 15, 1991.

Chilk, S. J., Secretary of the Commission. 1990a. NRC Memorandum for James M. Taylor, Executive Director for Operations. Subject: SECY-89-102 - Implementation of the Safety Goals. June 15, 1990

Chilk, S. J., Secretary of the Commission. 1990b. Nuclear Regulatory Commission Memorandum for James M. Taylor, Executive Director for Operations. Subject: SECY-90-16. Evolutionary Light Water Reactor (LWR) Certification Issues and Their Relationships to Current Regulatory Requirements. June 26, 1990.

Chilk, S. J., Secretary of the Commission. 1979. Nuclear Regulatory Commission Policy Statement, NRC Statement on Risk Assessment and the Reactor Safety Study Report (WASH-1400) in Light of the Risk Assessment Review Group Report. Attached to Memorandum for Lee V. Gossick, Executive Director for Operations. Subject: Staff Actions Regarding Risk Assessment Review Group Report. January 18, 1979.

Chung, K. M., and G. A. Hazelrigg. 1989. Nuclear Power Technology: A Mandate for Change. Nuclear Technology. 88(November).

Cohen, S. D., et al. 1990. Environmental Externalities: What State Regulators Are Doing. The Electricity Journal. July.

Cohen, S. 1989. Operating and Financial Risks in the Growing Capacity Shortage. A report prepared by Morgan Stanley. New York, N.Y. May.

Collier, H., Chairman, High-Level Radioactive Waste Disposal Committee. 1984. Report of Science Advisory Board. Letter to William Ruckleshaus. Report on Proposed Environmental Standards for the Management and Disposal of Spent Nuclear Fuel, High-Level and Transuranic Radioactive Wastes, 40 CFR 191. High-Level Radioactive Waste Disposal Subcommittee, Science Advisory Board, U.S. Environmental Protection Agency. February 17, 1984.

Committee on Suggested State Legislation, The Council of State Governments. 1991. Suggested State Legislation, Volume 50.

deBoer, C., and I. Catsburg. 1988. The Polls-A Report, the Impact of Nuclear Accidents on Attitudes Toward Nuclear Energy. Poll Report: Nuclear Energy, Public Opinion Quarterly. 52:254-261. American Association for Public Opinion Research. University of Chicago Press.

DOE. 1991. National Energy Strategy. First Edition, 1991/1992. Washington, D.C. February.

DOE, Energy Information Administration. 1990a. Annual Energy Outlook, Long Term Projections, 1990. DOE/EIA-0383(90). Released for printing January 12, 1990.

DOE. 1990b. Interim Report, National Energy Strategy, A Compilation of Public Comments. DOE/S-0066P. April.

DOE, Energy Information Administration. 1990c. Electric Plant Cost and Power Production Expenses 1988. DOE/EIA-0455(88). Released for printing August 16, 1990.

DOE. 1989a. Commercial Nuclear Power 1989, Prospects for the United States and the World. DOE/EIA-0438(89).

DOE. 1989b. Annual Outlook for U.S. Electric Power 1989, Projections Through 2000. DOE/EIA-047(89).

DOE, Energy Information Administration. 1989c. Annual Energy Review 1989. Energy Information Administration. DOE/EIA-0384(89). Released for printing May 24, 1990.

DOE, Energy Information Administration. 1989d. Nuclear Power Plant Construction Activity 1988. DOE/EIA-0473(88). Released for printing June 14, 1989.

DOE, Energy Information Adminstration. 1989e. Historical Plant Cost and Annual Production Expenses for Selected Electric Plants 1987. DOE/EIA-0455(87). Released for printing May 8, 1989.

DOE, Energy Information Adminstration. 1988c. An Analysis of Nuclear Power Plant Operating Costs. DOE/EIA-0511. Released for printing March 16, 1988; and letter dated August 10, 1990 from the Director of EIA's Office of Coal, Nuclear, Electric and Alternate Fuels to Archie L. Wood.

DOE. 1986a. Review of the Proposed Strategic National Plan for Civilian Nuclear Reactor Development. A Report of the Energy Research Advisory Board to the United States Department of Energy. DOE/S-0051. 1-4:(October).

DOE, Energy Information Adminstration. 1986b. An Analysis of Nuclear Power Plant Construction Costs. DOE/EIA-0485. March/April 1986.

DOE, Energy Information Administration. 1986c. Financial Analysis of Investor-Owned Electric Utilities. DOE/EIA-0499. November.

DOE, Energy Information Administration. 1982. Projected Costs of Electricity from Nuclear and Coal-Fired Power Plants. DOE/EIA-0356/2. 2(November).

EEI. 1989. Electricity Futures: America's Economic Imperative. January.

EEI Task Force on Nuclear Power. 1985. Report of the Edison Electric Institute on Nuclear Power. February.

Energy Daily. April 2, 1991. Appeals Court Reverses Earlier Ruling on NRC Licensing Rule. 19:62.

Environmental Protection Agency. 40 CFR Part 191.

EPRI. 1988. Status of Least-Cost Planning in the United States.

EPRI. 1986. Advanced Light Water Reactor Utility Requirements Document, Executive Summary, Part I. The Electric Power Research Institute Advanced Light Water Reactor Program. June.

Firebaugh, M. W., and M. J. Ohanian, eds. 1980. Gatlinburg II, An Acceptable Future Nuclear Energy System, Condensed Workshop Proceedings. Institute for Energy Analysis, Oak Ridge Associated Universities. March.

Flavin, C. 1988. The Case Against Reviving Nuclear Power. World-Watch. July - August. 1:4.

Fowler, T. K., and A. D. Rossin. 1990. First 1990 Group on Electricity. University of California, Berkeley. January 12, 1990.

Ganson, W. A., and A. Modigliani. 1989. Media Discourse and Public Opinion on Nuclear Power: A Constructionist Approach. Research supported by National Science Foundation grants SES-801642 and 8309343. University of Chicago. Reprints from Department of Sociology, Boston College.

George Washington University. 1989. International Conference on Enhanced Safety of Nuclear Reactors. August 9-10, 1988. Proceedings published as ITSR Report Number 008. Institute for Technology and Strategic Research, The School of Engineering and Applied Science. 192-193.

Giraud and Vendryes. 1989. Main Issues Requiring Resolution for Large-Scale Deployment of Nuclear Energy. International Workshop on the Safety of Nuclear Installations of the Next Generation and Beyond. Chicago. August 28-31, 1989.

Golay, M. W., and N. E. Todreas. 1990. Advanced Light-Water Reactors. Scientific American. April.

GSA. 1990. Title 10 (Energy) Code of Federal Regulations, Part 52. Published by the Office of the Federal Register, National Archives and Records Services, General Services Administration, as of January 1, 1990.

Hansen, Winje, Beckjord, et al. 1989. MIT Report, Making Nuclear Power Work: Lessons from Around the World. Technology Review, February/ March.

Harris, L. 1989. The Harris Poll, Sentiment Against Nuclear Power Plants Reaches Record High. Louis Harris and Associates. Released January 15, 1989.

IAEA. 1990. Nuclear Power Reactors in the World. Reference Data Series Number 2. Vienna, Austria. April.

INPO. 1989. Institute of Nuclear Power Operations, 1989 Annual Report. March.

Inside NRC. 1990. Outlook On State Regulation. April 9, 1990.

Inside NRC. 1989. Boston Edison Rate Settlement Makes Use of SALP Scores, INPO Indicators. An exclusive report on the U.S. Nuclear Regulatory Commission. McGraw-Hill. 11:22(October 23).

Jones, P. M. S., and G. Woite. 1990. Cost of nuclear and conventional baseload electricity generation. IAEA Bulletin. Quarterly Journal of the International Atomic Energy Agency. 32:3. Vienna, Austria.

Jordan, E., NRC. 1991. Facsimile dated March 19, 1991 to Theresa Fisher, National Research Council staff.

Lanouette, W. 1985. Nuclear Power in America. The Wilson Quarterly/Winter.

Lee Jr., B., President and Chief Executive Officer, Nuclear Management and Resources Council. 1991. Comments Letter to Samuel J. Chilk, Secretary, NRC. Subject: Notice of Availability, SECY-90-347 "Regulatory Impact Survey Report," 55 Fed. Reg. 53220 (December 27, 1990). January 28, 1991.

Lester, R. M., Driscoll, et al. 1985. National Strategies for Nuclear Power Reactor Development. Program on Nuclear Power Plant Innovation. MIT NPI-PA-002. March. (NSF Grant No. PRA 83-11777)

Lewis, H. W. 1986. Oversight Hearings. Testimony before the Subcommittee on Energy and the Environment of the Committee on Interior and Insular Affairs, U.S. House of Representatives, Ninety-Ninth Congress. June 10, 1986. Serial No. 99-68. (Lewis also provided testimony before this House of Representatives' Subcommittee on April 26, 1988 relating to creation of an independent Nuclear Safety Board. In addition, John F. Ahearne provided testimony relating to such a Board before the Subcommittee on Nuclear Regulation of the Senate Committee on Environment and Public Works on June 18, 1986, and Bill S.14 was introduced in the Senate on January 6, 1987 to amend the Energy Reorganization Act of 1974 to create an independent Nuclear Safety Board.)

Lewis, H. W., Chairman. 1978. Risk Assessment Review Group Report to the U.S. Nuclear Regulatory Commission. NUREG/CR-0400. September.

Moynet, G., et al. 1988. Electricity Generation Costs Assessment Made in 1987 for Stations to be Commissioned in 1995. UNIPEDE (International Union of Producers and Distributors of Electrical Energy). Sorrento Congress. May 30-June 3, 1988.

Myers, R. 1990. Nuclear Industry, Sitting Pretty. U.S. Council for Energy Awareness. Summer.

National Association of Regulatory Utility Commissioners. 1988. Least-Cost Utility Planning: A Handbook for Public Utility Commissioners, Volumes 1 & 2.

National Independent Energy Producers. 1991. Written Statement before the U.S. Senate Committee on Energy and Natural Resources. February 21, 1991.

National Independent Energy Producers. 1990. Bidding for Power: The Emergence of Competitive Bidding in Electric Generation. March.

National Research Council. 1990. Rethinking High-Level Radioactive Waste Disposal. A Position Statement of the Board on Radioactive Waste Management. National Academy Press. July.

Nealy, S. M. 1990. Nuclear Power Development, Prospects in the 1990s. Battelle Press. Columbus, Ohio.

North American Electric Reliability Council. 1991. Electricity Supply & Demand 1991-2000. July.

North American Electric Reliability Council. 1990. 1990 Electricity Supply & Demand for 1990-1999. November.

North American Electric Reliability Council. 1989. 1989 Electricity Supply & Demand for 1989-1998. October.

North American Electric Reliability Council. 1988. 1988 Electricity Supply & Demand for 1988-1997. October.

North American Electric Reliability Council. 1987. 1987 Electricity Supply & Demand for 1987-1996. November.

North American Electric Reliability Council. 1986. 1986 Electricity Supply & Demand for 1986-1995. October.

NPOC. 1990. A Perfect Match: Nuclear Energy and The National Energy Strategy. A Position Paper by the Nuclear Power Oversight Committee. November.

NRC. 1991a. Nuclear Regulatory Commission Information Digest, 1991 Edition. NUREG-1350. 3(March).

NRC. 1991b. Preliminary Notification of Event or Unusual Occurrence. PNO-IIT-91-02A. August 22, 1991.

NRC. 1991c. Preliminary Notification of Event or Unusual Occurrence. PNO-IIT-91-02. August 19, 1991.

NRC. 1991d. Policy Statement, Possible Safety Impacts of Economic Performance Incentives. 7590-01. July 18, 1991.

NRC. 1990a. Nuclear Regulatory Commission Information Digest, 1990 Edition. NUREG-1350. 2(March).

NRC. 1990b. Licensed Operating Reactors, Status Summary Report. Data as of 12-31-89. NUREG-0020. 14:1(January).

NRC. 1990c. 10 CFR Part 51. Consideration of Environmental Impacts of Temporary Storage of Spent Fuel After Cessation of Reactor Operation; and Waste Confidence Decision Review. Final Rules, Federal Register. 55:181(September 18).

NRC. 1990d. Survey of NRC Staff Insights on Regulatory Impact. SECY-90-250, July 16, 1990.

NRC. 1990e. Industry Perceptions of the Impact of the U.S. Nuclear Regulatory Commission on Nuclear Power Plant Activities. NUREG-1395. Draft report. March.

NRC. 1990f. Licensed Operating Reactors Status Summary Report (Data as of 12/31/89). NUREG-0020. 14:1(February).

NRC, Advisory Committee on Reactor Safeguards. 1990g. Letter to Chairman Carr. Subject: Review of NUREG-1150, "Severe Accident Risks: An Assessment for Five U.S. Nuclear Power Plants." November 15, 1990.

NRC. 1989a. Information Digest, 1989 Edition. NUREG-1350. 1(March).

NRC. 1989b. Office for Analysis and Evaluation of Operational Data. 1989 Annual Report. Power Reactors. NUREG-1272. 4:1.

NRC. 1989c. Severe Accident Risks: An Assessment for Five U.S. Nuclear Power Plants. NUREG-1150. 1(June).

NRC. 1989d. Office for Analysis and Evaluation of Operational Data, 1989 Annual Report. Power Reactors. NUREG-1272. 4:1.

NRC, Advisory Committee on Reactor Safeguards. 1988a. Letter to Chairman Kenneth M. Carr, Subject: Program to Implement the Safety Goal Policy -- ACRS Comments. April 12, 1988.

NRC. 1988b. Generic Letter No. 88-20, Individual Plant Examination for Severe Accident Vulnerabilities - 10 CFR 50.54(f). November 23, 1988.

NRC, Advisory Committee on Reactor Safeguards. 1987a. Letter to Chairman Kenneth M. Carr, Subject: ACRS Comments on an Implementation Plan for the Safety Goal Policy. May 13, 1987.

NRC. 1982. Nuclear Power Plants Construction Status Report (Data as of 6-30-82). NUREG-0030. 6:2(October).

Nuclear Power Oversight Committee. 1991. Position Paper on Standardization. JJT/634P. January 9, 1991.

Nuclear Power Oversight Committee. 1986. Leadership in Achieving Operational Excellence--The Challenge for all Nuclear Utilities. Chicago, Illinois. August. Called the "Sillin Report" (Lee Sillin chaired the report committee).

Nuclear Waste Technical Review Board. 1990. First Report to the U.S. Congress and the Secretary of Energy. p. 31. March.

OECD Nuclear Energy Agency/International Energy Agency. 1989. Projected Costs of Generating Electricity from Power Stations for Commissioning in the Period 1995-2000. Paris.

Pasternak, A. D., and R.J. Budnitz. 1987. State-Federal Interactions in Nuclear Regulation. UCRL-21090. S/C 5221201. Lawrence Livermore National Laboratory. December.

Presidential Commission on Catastrophic Nuclear Accidents. 1990. Report to the Congress. Volume One. August.

Price-Anderson Amendments Act of 1988. Public Law 100-408, 102 Statute 1066. August 20, 1988.

Sagan, C. 1990. Tomorrow's Energy, How to Have Your Cake and Eat It Too. Parade Magazine. November 25, 1990.

Seismicity Owners Group and Electric Power Research Institute. 1986. Seismic Hazard Methodology for the Central and Eastern United States. EPRI NP-4726. July.

Stello, Jr., V., Executive Director for Operations, U.S. Nuclear Regulatory Commission. 1989. Memorandum for the Commissioners. Subject: Implementation of Safety Goal Policy. SECY-89-102. March 30, 1989.

Union of Concerned Scientists. 1987. Safety Second. The NRC and America's Nuclear Power Plants. Indiana University Press.

U.S. Congress. Office of Technology Assessment. 1984. Nuclear Power in an Age of Uncertainty. OTA-E-216. February.

U.S. Congress. 1981. Nuclear Regulatory Legislation Through The Ninety Sixth Congress, Second Session. Prepared for the Committee on Environment and Public Works. 97th Congress. 1st Session. Committee Print. Serial Number 97-3. Atomic Energy Act of 1954, Section 2e. August.

U.S. Congress. 1978. United States Code, Congressional and Administrative News, 95th Congress - Second Session. Volume 2. Public Law 95-617 (H.R. 4018). November 9, 1978. Public Utility Regulatory Policies Act of 1978.

U.S. Court of Appeals for the District of Columbia Circuit. 1990. Opinion on Petition for Review of An Order of the Nuclear Regulatory Commission. No. 89-138. Decided November 2, 1990.

U.S. General Accounting Office. 1990. Electricity Supply, The Effects of Competitive Power Purchases Are Not Yet Certain. Report to the Chairman of the Subcommittee on Oversight and Investigations, Committee on Energy and Commerce, House of Representatives. GAO/RCED-90-182. August.

Weinberg, A. M. 1989. Engineering in an Age of Anxiety: The Search for Inherent Safety. Engineering and Human Welfare Symposium Program and Papers. National Academy of Engineering 25th Annual Meeting. October 4, 1989.

Willrich, M. 1985. Nuclear Power in a Changing U.S. Electric Utility Industry. Information of Interest from Pacific Gas and Electric Company Corporate Communications.

10 CFR Part 50, Safety Goals for the Operations of Nuclear Power Plants. Policy Statement. Republication in Federal Register, PS-PR-51. November 30, 1988.

NOTE: The following goals, objectives, and proposed guideline are contained in the above reference.

This policy statement contains two qualitative safety goals that are supported by two quantitative objectives. It also contains a general performance guideline.

Qualitative Safety Goals:

Individual members of the public should be provided a level of protection from the consequences of nuclear power plant operation such that individuals bear no significant additional risk to life and health.

Societal risks to life and health from nuclear power plant operation should be comparable to or less than the risks of generating electricity by viable competing technologies and should not be a significant addition to other societal risks.

Quantitative Objectives:

The risk to an average individual in the vicinity of a nuclear power plant of prompt fatalities that might result from reactor accidents should not exceed one-tenth of one percent (0.1 percent) of the sum of prompt fatality risks resulting from other accidents to which members of the U.S. population are generally exposed.

The risk to the population in the area near a nuclear power plant of cancer fatalities that might result from nuclear power plant operation should not exceed one-tenth of one percent (0.1 percent) of the sum of cancer fatality risks resulting from all other causes.

The Commission proposed for further staff examination the following general performance guideline.

Consistent with the traditional defense-in-depth approach and the accident mitigation philosophy requiring reliable performance of containment systems, the overall mean frequency of a large release of radioactive materials to the environment from a reactor accident should be less than 1 in 1,000,000 per year of reactor operation.

The narratives describing the various technologies are based upon the oral and written record submitted to the Committee. These summaries represent a conscientious effort to accurately depict the nature, attributes, and distinguishing features of each technology. The Committee does not represent this Chapter as a comprehensive treatment of each advanced nuclear reactor technology or as an independent verification of all vendor representations.

3

Assessment of Advanced Nuclear Reactor Technologies

The Committee was asked to perform a critical comparative analysis of the practical technological options for the future development of nuclear power. In conducting this analysis the Committee undertook the following tasks:

- identifying the full range of practical nuclear reactor technologies for the next generation of nuclear plants;
- developing criteria to evaluate these technologies; and
- evaluating the technologies in terms of the criteria developed.

The Committee developed evaluation criteria that reflected the characteristics deemed most important for future nuclear power plants (e.g., safety and cost). (see Appendix B) The Committee then invited reactor vendors to present design concepts for advanced nuclear reactor technologies. Enhanced and novel features of these technologies are first described, and then the technologies are evaluated in light of the Committee's criteria.

OVERVIEW OF ADVANCED REACTOR TECHNOLOGIES

Most reactors operate by fissioning uranium atoms with slow or thermal neutrons. Thermal neutrons are produced in moderators such as graphite or water. The reactor cores are usually cooled with water or a gas (e.g., helium). Some reactors have no moderator, operate with fast neutrons, and are normally cooled by a liquid metal (e.g., sodium). A summary of the advanced reactor technologies reviewed by the Committee is given in Table 3-1, based on vendor-provided information. The major headings in Table 3-1 (Large Evolutionary Light Water Reactors, etc.) align with the titles of the major sections below in which the advanced reactors are discussed. The acronyns in Table 3-1 are explained in the following paragraphs.

TABLE 3-1 Vendor Descriptions of Technical Aspects of Advanced Nuclear Reactors

Reactor Designation[a]	Vendor	Power (MWe)	Passive Containment Cooling	Passive Residual Heat Removal	Passive Emergency Core Cooling System	Primary Coolant	Digital Control
Large Evolutionary Light Water Reactors							
ABWR	GE [GE Nuclear Energy, 1989]	1,350	No	No	No	Water	Yes
APWR-1300[b]	Westinghouse [McCutchan et al., 1989]	1,350	No	No	No	Borated water	Yes
System 80+ PWR	CE[c] [CE, 1989a]	1,300	Being Evaluated	No	No	Borated water	Yes
Mid-Sized Light Water Reactors With Passive Safety Features							
AP-600 PWR	Westinghouse [Westinghouse, 1989]	615	Yes	Yes	Yes	Borated water	Yes
SBWR	GE [GE Nuclear Energy, 1989]	600	Yes	Yes	Yes	Water	Yes
Other Reactor Concepts							
CANDU 3 HWR	AECL [AECL, Undated; AECL, 1989]	450	No	Yes[i]	Yes/No[e]	Heavy water	Yes, with automated startup
SIR PWR	CE[c] [CE, 1989b]	320[h]	Yes	Yes	Yes/No[e]	Water	Yes
MHTGR	GA [GA, 1989]	134[f]	Yes	Yes[d]	Yes[d]	Helium	Yes, with automated startup
PIUS PWR	ABB Atom [ABB, 1989]	640	Yes	Yes[d]	Yes[d]	Borated water	Yes
PRISM LMR	GE [Berglund, 1989; Till, 1989]	155[g]	Yes	Yes[d]	Yes[d]	Sodium	Yes, with automated startup

a All designs include load-following capability of between 50 and 100 percent.
b Westinghouse also supplied information on the APWR-1000, a 1,050 MWe plant whose features are similar to the APWR-1300, except for the core design.
c Combustion Engineering; now Asea Brown Boveri Combustion Engineering Nuclear Power.
d The residual heat removal system and the emergency core cooling system are essentially the same system.
e Yes at high pressure; no at low pressure.
f The power plant design includes four 134 MWe reactor modules for a total of 536 MWe.
g The power plant design includes one to three power blocks, each containing three 155 MWe reactor modules for a total of 465 MWe per block, and a net electrical plant rating up to 1,395 MWe.
h The power plant design includes two 320 MWe reactor modules on one turbine generator to produce 640 MWe output.
i System under development.

The advanced commercial water reactors reviewed are of three classes: (1) pressurized water reactors (PWR) are light water reactors (LWR) that maintain the water adjacent to the fuel elements at high pressure to prevent boiling; (2) boiling water reactors (BWR) are LWRs in which the water adjacent to the fuel elements boils; and (3) heavy water reactors (HWR) are reactors in which heavy water (deuterium oxide or D_2O) serves as both coolant and moderator instead of ordinary (light) water, and only the coolant is pressurized. In current HWRs the reactor fuel is natural uranium, and in LWRs the fuel is uranium enriched to contain up to a few percent of the uranium-235 isotope. (APWR and ABWR mean "advanced"; AP means "advanced passive"; SBWR means "simplified"; CANDU means "Canadian deuterium uranium"; SIR means "safe integral reactor"; and PIUS means "process inherent ultimate safety".)

Two other advanced reactor technologies reviewed by the Committee do not use water as a coolant or moderator. They are the gas-cooled graphite-moderated reactor known as the MHTGR (modular high-temperature gas-cooled reactor) and the liquid metal-cooled fast neutron reactor known as the PRISM LMR (power reactor, innovative small module liquid metal reactor).

The vendors, in order of appearance in Table 3-1, are General Electric (GE), Westinghouse, Combustion Engineering (CE), Atomic Energy of Canada Limited (AECL), General Atomics (GA), and Asea Brown Boveri (ABB) Atom.

The following sections treat ten advanced reactor types--three large evolutionary LWRs, two mid-sized LWRs with passive safety features, and five other reactor concepts.

Large Evolutionary Light Water Reactors

Evolutionary LWRs, a subset of advanced reactors consisting of the ABWR, APWR-1300, and System 80+, are improved versions of current LWRs with capacities of greater than 1,000 megawatts electric (MWe). These evolutionary designs differ to some extent from current LWRs, for which thousands of reactor years of operating experience have been accumulated worldwide. All evolutionary designs seek greater safety margins, greater ease of construction, improved reliability and availability, improved maintainability, lower costs, and greater ease of operation over existing large LWRs. The evolutionary reactor designs conform to the advanced LWR requirements contained in the Utility Requirements Document.[EPRI, 1990] A summary of these requirements, which cover both enhanced safety and improved economics, is presented in Table 3-2. The Utility Requirements Document is being prepared through the Electric Power Research Institute (EPRI). The technical judgments on all significant issues are reviewed by a Utility Steering

TABLE 3-2 Key Utility Design Requirements for Advanced Light Water Reactors**

Plant size	Reference size 1,200-1,300 MWe for evolutionary designs; reference size 600 MWe for passive safety designs
Design life	60 years
Design philosophy	Simple, rugged, no prototype required
Accident resistance	≥ 15 percent fuel thermal margin, increased time for response to upsets
Core damage frequency	< 10^{-5}/year by probabilistic risk analysis
Loss of coolant accident	No fuel damage for 6" pipe break
Severe accident mitigation	< 25 REM at site boundary for accidents with > 10^{-6}/year cumulative frequency
Emergency planning zone	For passive plant provide technical basis for simplification of off-site emergency plan
Design availability	87 percent
Refueling interval	24 months capability
Maneuvering	Daily load follow
Worker radiation exposure	< 100 person REM/year
Construction time	1,300 MWe: ≤ 54 months (first concrete to commercial operation); 600 MWe: ≤ 42 months
Design status	90 percent complete at construction initiation
Economic goals	10 percent cost advantage over alternatives (nonnuclear) after 10 years and 20 percent advantage after 30 years
Resulting cost goals (1989 $)	

	1,200 MWe*	600 MWe*
Overnight capital	1,300 $/kWe	1,475 $/kWe
30-year levelized total generation	6.3 cents/kWh	7.2 cents/kWh

*1,200 MWe commercial operation in 1998; 600 MWe in 2000

SOURCE: Electric Power Research Institute. Advanced Light Water Reactor Utility Requirements Document, Volume 1, ALWR Policy and Summary of Top-Tier Requirements. Issued 3/90. Palo Alto, California.

** These requirements apply to both the large evolutionary LWRs and to the mid-sized LWRs with passive safety features.

Committee made up of experienced nuclear utility executives from throughout the United States and abroad.

The first standardized design to be certified in the United States is likely to be an evolutionary LWR. Three of these LWR design concepts were presented to the Committee. Only the new or unique features of each concept will be described.

Advanced Boiling Water Reactor

The 1,350 MWe ABWR is being developed as the next Japanese standard BWR under the leadership of the Tokyo Electric Power Company in a joint venture with GE, Hitachi, Toshiba, and a group of Japanese utilities. In 1989 the Tokyo Electric Power Company announced its decision to proceed with the construction of two ABWR units at its Kashiwazaki-Kariwa Nuclear Power Station, with commercial operation of the first unit scheduled for 1996 and of the second for 1998. GE was selected to supply the nuclear steam supply systems, fuel, and turbine generators. Figure 3-1 is a diagram of this advanced reactor's pressure vessel and core.

Finally, GE has applied for design certification under 10 CFR Part 52, and certification currently is scheduled for completion in the mid-1990s. GE expects that this reactor will be the first certified U.S. standard plant.[Wolfe and Wilkens, 1988]

Core Design. A new core and fuel design has been developed to increase operating economies, and external recirculation pumps have been replaced by internal pumps. The reactor pressure vessel has a single forged ring for the 10 internal pump nozzles and the conical support skirt. The elimination of the external recirculation pump piping and the use of the vessel forged rings have resulted in a 50 percent reduction in the weld requirements for the primary system pressure boundary. Finally, the reactor pressure vessel is standard BWR design, except that (1) the annular space between the pressure vessel shroud and the vessel wall is increased, and (2) the standard cylindrical vessel support is now a conical skirt.

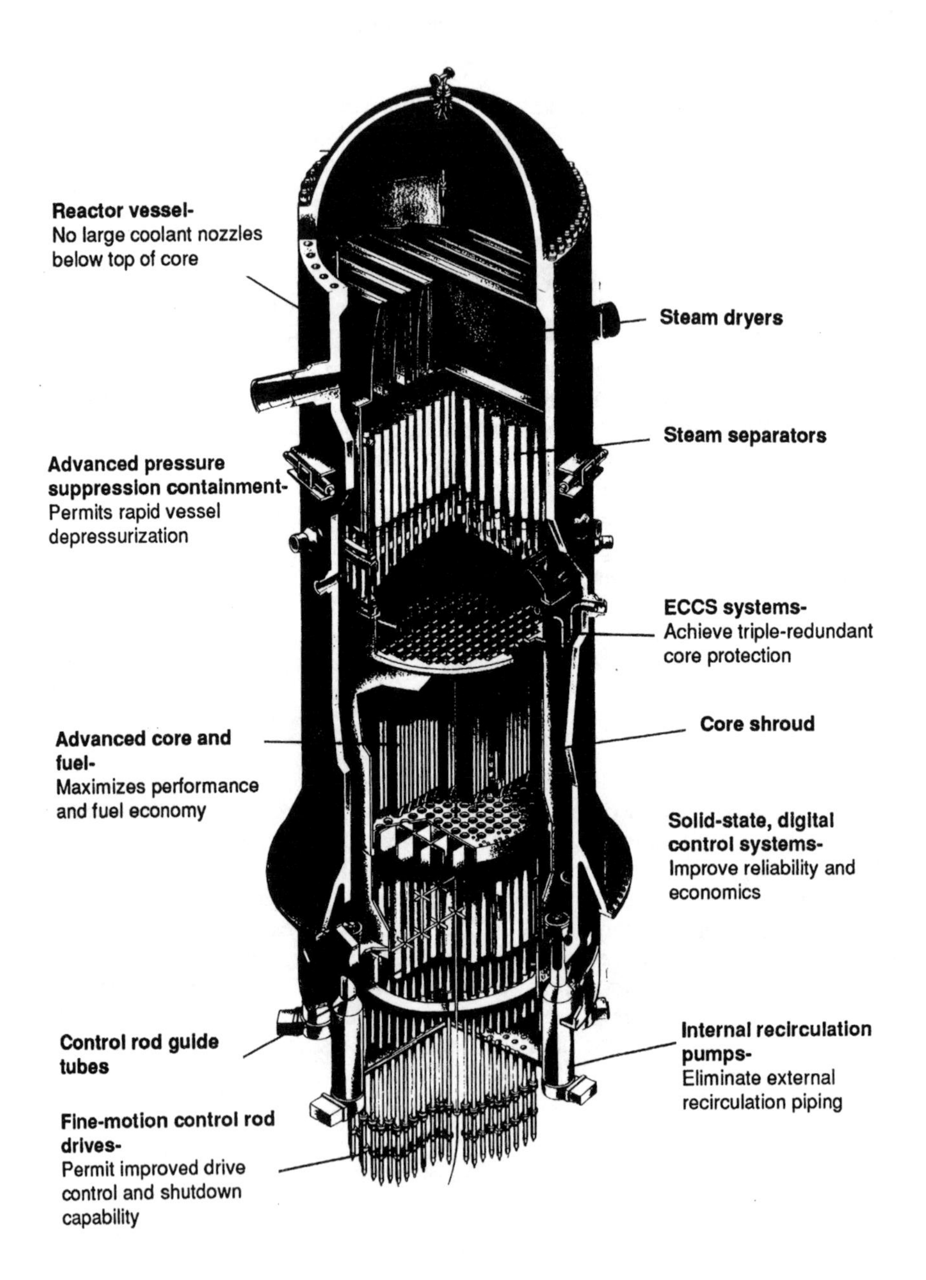

FIGURE 3-1 Advanced boiling water reactor, pressure vessel and core. SOURCE: GE Nuclear Energy

Fluid Systems. The emergency core cooling system and residual heat removal system have a three-division scheme. Two divisions each provide both high-pressure and low-pressure emergency core coolant injection capability. The third division combines a reactor-steam-driven turbine pump for the high-pressure coolant injection and low-pressure coolant injection system. The steam driven system is the conventional reactor core isolation cooling (RCIC) system that has been upgraded to a safety system. The other two divisions are the high-pressure core flooders. The steam driven system is controlled by water level and is the first high-pressure system to come on in the event of a loss-of-coolant accident (LOCA) or a reactor isolation transient. The residual heat removal system is a triply redundant water delivery/decay heat removal combination. Additionally, the elimination of large nozzles on the reactor vessel below the core helps ensure that the core is not uncovered during any LOCA. At the same time, a 50 percent reduction of the total required emergency core cooling system pumping capacity is realized, compared to an equivalent-size external loop BWR plant.

Control and Instrumentation. The control and instrumentation system features a multiplexing system that complements a digital, solid-state control design. This equipment permits a design that increases the system redundancy, provides fault-tolerant operation, and provides self-diagnostics while the system is in operation.[1]

Containment. The reactor building/containment is a steel-lined reinforced concrete structure with a covered pressure suppression pool. The design also features a horizontal vent system for venting the drywell to the suppression pool in the event of a LOCA. In addition, elimination of the external recirculation piping system permits greater access for inspection and maintenance of the drywell.

---

[1] Multiplexing will be considered, as will all the advanced instrumentation and controls technology, as part of the licensing process for the large evolutionary reactors. This will establish the precedent for the other advanced reactors. Included in the licensing review will be digital controls technology and the new control room designs that incorporate current human factors considerations.[M. Chiramal, Section Chief, Advanced Reactor Section, Instrumentation and Control, U.S. Nuclear Regulatory Commission, personal communication, August 29, 1991.]

Advanced Pressurized Water Reactor

The design for a large evolutionary APWR has been developed by Westinghouse in cooperation with five Japanese utilities and Mitsubishi. Kansai Electric Company has declared its intention to build the first such plant, pending approval of a suitable site.[Hirata et al., 1989] This four-loop 1,350 MWe model incorporates several technological advancements.[McCutchan et al., 1989] Although it was primarily developed for Japan, the design concepts were adopted in the criteria specified by EPRI. Figure 3-2 depicts the reactor's integrated safety systems.

Core Design. The most significant new feature of the APWR is the 15 to 20 percent reduction in power density for greater safety and thermal operating margins. Reactivity is controlled with rods that displace water in the lattice during the first part of the refueling cycle; the water is returned later in the cycle by removing the displacement rods. (This feature is not included in the APWR-1000 design, which has a conventional but reduced power density core.) It is claimed that these features combine to reduce fuel costs by 20 percent. In addition, the increase in the number of movable rods compared to conventional designs requires a larger rod-guide region above the core. The larger reactor vessel provides an increased inventory of cooling water above the core, leading to enhanced safety while reducing requirements for the emergency core cooling system (ECCS).

Steam Generators. The U-tube steam generators are larger than those in existing Westinghouse reactors, with lower average temperatures, lower heat flux, and easier accessibility for maintenance and repair. Other features include improved tube materials and an improved tube support plate design.

Fluid Systems. Safety and control functions have been integrated, reducing piping requirements and enhancing safety-related fluid system design. For the ECCS, four high-pressure pumps take suction from an in-containment refueling water storage tank and inject borated water into the reactor vessel to improve core protection for small pipe breaks. This eliminates the switchover from a tank located outside the containment to a sump inside the containment.

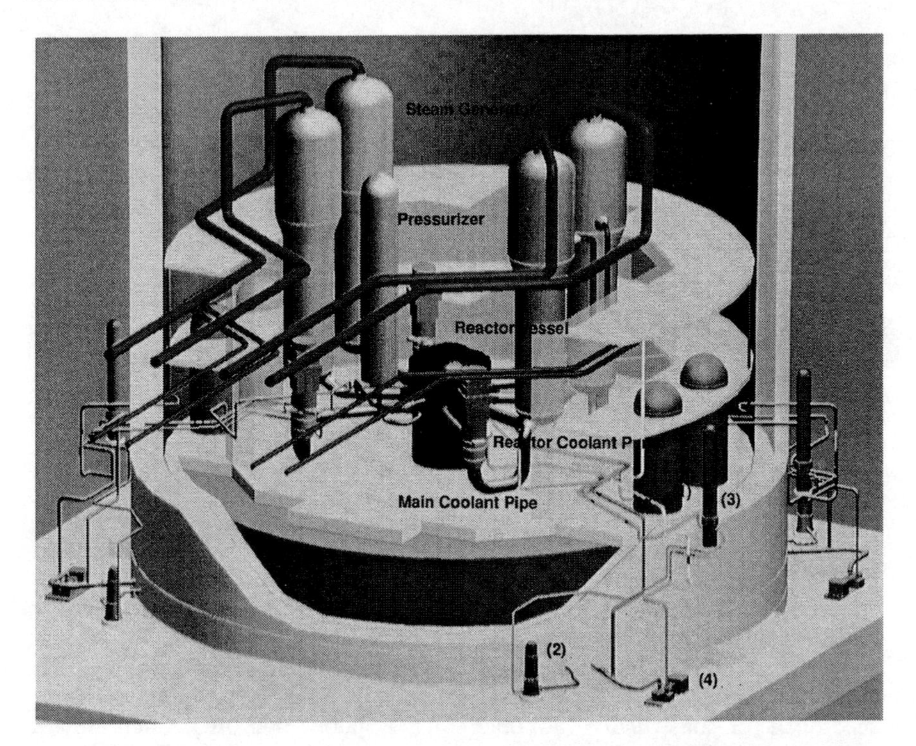

FIGURE 3-2 Advanced pressurized water reactor integrated safety systems (1 = Accumulator; 2 = High head safety injection pump; 3 = Residual heat removal heat exchanger; 4 = Residual heat removal/coolant systems pump). SOURCE: Westinghouse Energy Systems

Control and Instrumentation. The integrated control safety systems feature microprocessors and multiplexed data highways that allow complete and rapid communication between the central control room and the various control and protection points in the plant. The multiplexed interconnections reduce control cabling by up to 70 percent. The safety system is designed to operate automatically when plant conditions reach trip set points.

Containment. A double cylindrical containment building is used with an interior pressure bearing steel shell and an external concrete shield wall. The steel containment shell is easier to construct to quality standards. The total containment volume is increased, and congested areas have been eliminated.

System 80+ Standard Design Pressurized Water Reactor

The System 80+ PWR, the third large evolutionary reactor reviewed by the Committee, is rated at 1,300 MWe. It is the result of a design effort led by CE (now Asea Brown Boveri Combustion Engineering Nuclear Power), assisted by the Duke Power Company and the Korea Advanced Energy Research Institute. This design evolved from CE's System 80 nuclear steam supply system design. The advanced System 80+ design draws heavily on the designs of three operating System 80 units at Palo Verde and two more scheduled for construction in Yonggwang, Republic of Korea. Incremental improvements to the components that are currently used have been incorporated in the new design.[CE, 1989a] Figure 3-3 is an elevation view of the System 80+ containment building.

Core Design. The System 80+ core design uses only control rods for reactivity control, thus eliminating the need to adjust the boron concentration in the coolant. This feature simplifies reactivity control during power load changes. In addition, the core thermal operating margin has been increased by reducing normal operating hot leg temperatures and revising monitoring methods.

Steam Generators. Design enhancements in the steam generators include better steam dryers, an increased overall heat transfer area, and slightly reduced full power steam pressure resulting from lower coolant temperatures, compared to the System 80 design. Additional heat transfer area permits the nuclear steam supply system to maintain rated output with a significant

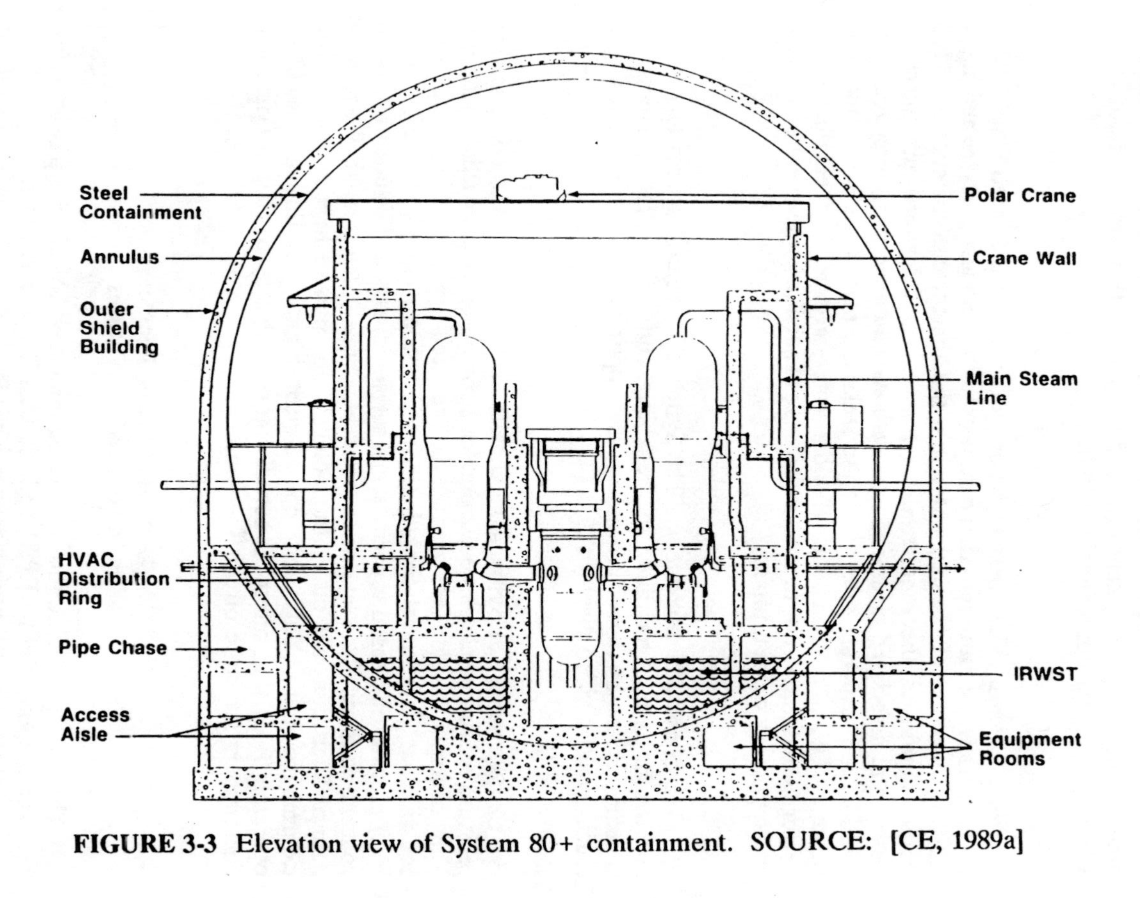

FIGURE 3-3 Elevation view of System 80+ containment. SOURCE: [CE, 1989a]

number of tubes plugged. Each steam generator will also have a larger secondary feedwater inventory which extends the "boil dry" time, enhancing the nuclear steam supply system's capability to tolerate upset conditions and thereby improving operational reliability.

Fluid Systems. The safety injection system in the enhanced System 80+ is a four-train system of injection pumps used for both low-pressure and high-pressure injection of borated water into the reactor coolant system. This feature eliminates the requirement for a dedicated low-pressure injection system and associated cross-connects with the shutdown cooling system. In addition, four separate safety injection tanks are part of the safety injection system. The in-containment refueling water storage tank eliminates the reliance on automatic or manual switchover of suction in the event of a break in the primary coolant piping.

Control and Instrumentation. The System 80+ control system features a new design to meet human factor, reliability, and licensing requirements. It is characterized by digital processing, fiber optic data communications, and touch-sensitive video displays.

Containment. The System 80+ containment design is a 200-foot-diameter pressure bearing steel sphere surrounded by an outer concrete shield building. The concrete shield that surrounds the steel sphere offers secondary containment, and the relatively large free internal volume (3.4 million cubic feet) provides increased capacity for absorbing energy and diluting hydrogen concentrations in the event of an accident. Finally, the steel shell acts as a natural heat sink and offers the potential for passive heat removal using external cooling. This steel containment building is designed with an operating floor that offers 75 percent more usable space than a cylindrical containment structure of equal volume.

Mid-Sized Light Water Reactors With Passive Safety Features

The principal U.S. effort to develop mid-sized LWRs with passive safety features is sponsored by EPRI and the U.S. Department of Energy (DOE), with substantial contributions from major U.S. suppliers.[Taylor and Stahlkopf, 1988; Taylor, 1989] (EPRI receives funding from most U.S. utilities and utilities in France, Italy, the Netherlands, Japan, South Korea, and Taiwan.) The passive plant was envisioned as a smaller reactor that would employ primarily passive means--gravity, natural circulation, and stored energy--for its essential safety functions.

The passive LWR design concept was considered potentially attractive to utility investors for several reasons: (1) the fundamental simplicity of the passive safety concept offers an opportunity to effect wholesale simplification (reducing many valves, pumps, pipes, tanks, instruments, etc.), with attendant improvement in construction costs and schedules, and plant operability and maintainability; and (2) by reducing reliance on active components and human intervention, passive features can help accommodate a wide range of upset conditions and internal and external plant threats, such as loss of all electrical power.[Westinghouse, 1989]

It is estimated that, compared to a conventional 600 MWe pressurized LWR, a plant with passive cooling features would offer the following savings in bulk commodities:

- 60 percent fewer valves;
- 35 percent fewer large pumps;
- 75 percent less piping (in the nuclear island, the predominantly nuclear part of the plant);
- 80 percent less heating, ventilation, and air-conditioning ducting;
- 80 percent less control cable (nuclear island); and
- 50 percent less seismic building volume.[Westinghouse, 1989; Taylor and Stahlkopf, 1988]

For a BWR, the following reductions would be achieved:

- valves by 16 percent;
- safety-grade pumps and valves by 26 percent;
- fans by 80 percent; and
- large pumps by 73 percent.[Taylor, 1989]

The Chairman of the Utility Steering Committee for EPRI's Advanced LWR Program provided the following thoughts on the choice of the 600 MWe size:

> This choice was more or less arbitrary. It was arrived at from two directions. The first was that, in discussions with utilities before the ALWR [Advanced LWR] Program began, EPRI concluded that there were a number who felt a smaller size plant in the approximately 600 MWe size range would be better adapted to their system, and would be something more easily accepted, than a plant twice that size. The second reason for the choice was to distance the Passive Plant from the Evolutionary Plant so as to reduce the direct competition between the two.

Enough work has been done to say that 600 MWe is not the limit for gravity removal of decay heat. Both General Electric and Westinghouse have done work, sometimes with Japanese firms, which indicates that plants 900 to 1,000 MWe are feasible. On the other hand, we have not done enough design or experimental work in the United States to say with confidence where a limit rests.

It is true that there is some value established by the size of the reactor vessel or the size of containment which prudently limits the capacity of the first generation of Passive Plants. That is because the power densities are lower and therefore the core size is larger for a given capacity.

I might say that many of us believe this to be an advantage in an overall sense in that we believe one of the problems with present generation plants is that sizes and power densities were pushed too far, too quickly.[Kintner, 1989]

Advanced Passive Pressurized Water Reactor

The advanced passive (AP-600) design was developed by Westinghouse with financial support from DOE and EPRI. Figure 3-4 is a diagram of the AP-600 passive cooling system.[Westinghouse, 1989]

Core Design. The AP-600 has the proven uranium oxide fueled core, with reductions in coolant temperature, flow rates, and core power density to increase design thermal margins.

Steam Generators. The steam generators, of U-tube design, include evolutionary improvements over those in existing plants, including improved tube material to reduce corrosion and upgraded antivibration bars to reduce wear. Lower average coolant temperatures are intended to improve tube integrity. The reactor coolant pumps are mounted in the channel head at the bottom of the steam generator, simplifying the support system, reducing piping and construction, and increasing the space for maintenance.

Fluid Systems. Passive cooling in the AP-600 is achieved with a passive ECCS, which is a combination of two cooling water sources: (1) gravity drain of water from two core makeup tanks and (2) a large refueling water storage tank suspended above the level of the core. Additionally, the ability to inject water from two pressurized accumulator tanks is retained. Core decay heat can also be removed through a passive residual heat exchanger located in the

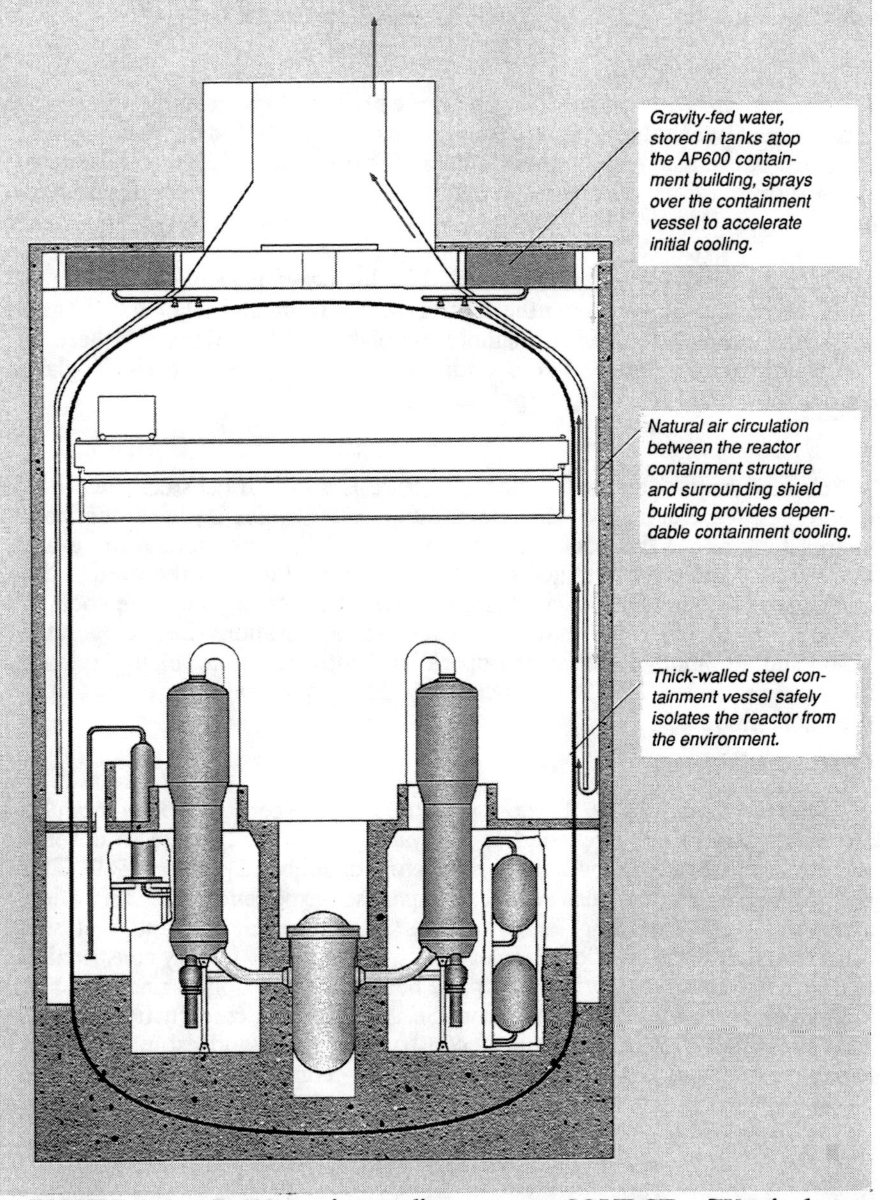

FIGURE 3-4 AP-600 passive cooling systems. SOURCE: [Westinghouse, 1989]

refueling water storage tank. This heat exchanger transfers decay heat to the refueling water by natural circulation as shown in Figure 3-4.

Control and Instrumentation. Features of the system include microprocessor-based technology, multiplexed controls for plant data and signals, electrical data links, and fiber-optic data highways. It includes an advanced alarm system, advanced operation display system, and an advanced accident monitoring/safety display system.

Microprocessors and multiplexed data highways permit complete and rapid communication between the central control room and other control and protection cabinets located throughout the plant. Malfunctions anywhere in the plant can be detected and addressed on a real-time basis if plant conditions change from trip setpoints.

Containment. The containment structure is a cylindrical steel shell that, in emergencies, can be cooled by evaporating water, which is gravity-fed from a large tank above the containment structure. This tank holds a three-day water supply and can be refilled externally. Heat is ultimately removed to the atmosphere by a natural air circulation system. Like emergency core cooling, containment cooling requires only automatic valve operations (i.e., no operator action and no pump, diesel, or fan operations) after any major energy release from the maximum LOCA. Concrete shielding is provided external to the steel containment.

Modular Construction. Large-scale studies on the construction of modules are being carried out by Avondale Shipyards and Westinghouse to develop economical assembly techniques in the factory or shipyard.[Taylor, 1989] This construction planning also reflects Japanese experience in fabricating, assembling, and installing large modules in their nuclear plants. It was estimated that these smaller, simpler plants, amenable to factory construction and with the design essentially complete before construction begins, could be built in three to four years following the issuance of a construction permit. This simplified design with shorter construction times and estimated lower capital costs could compensate for the loss of economy of scale credited to larger plants.

Simplified Boiling Water Reactor

The SBWR is a passive design being developed by GE with financial support from DOE and EPRI.[Duncan and McCandless, 1988] Figure 3-5 illustrates this reactor concept.

Core Design. The SBWR's lower power density increases thermal margins in the critical power ratio from 10 percent to more than 30 percent. This indicates that the power at the transition from nucleate to film boiling, relative to the operating power, has increased by 20 percent. The reactor operates at full power with natural circulation of water so that the recirculation pumps are eliminated, resulting in a simpler reactor vessel, reduced vulnerability to loss of coolant, and reduced maintenance. The larger reactor vessel needed for natural circulation provides the additional benefit of a greater inventory of water above the core at the initiation of any transient conditions.

Fluid Systems. Passive cooling is achieved by locating the suppression pool above the reactor core so that, in an emergency, core cooling is achieved by gravity rather than safety injection pumps. This feature not only eliminates the injection pumps, but also associated valves, piping, and diesel generator power supplies. The suppression pool is a standard feature of current BWRs. It serves as a passive cooling system that reduces the temperature and pressure in the containment building in the event of a severe accident.

During normal operation, an isolation condenser submerged in a pool of water, located above the core and outside the containment, controls reactor pressure passively (automatically) without reducing the fluid volume in the reactor vessel. Isolation condensers for passive reactor pressure control were used in early BWRs and have been reintroduced in this design. These isolation condensers can also be used to remove long-term, postaccident decay heat from the containment. This second passive feature would function in the event of loss of coolant.

Control and Instrumentation. The system includes an advanced control panel design and features an intelligent multiplexing system using fiber optic data transmission and extensive use of standard microprocessor-based control and instrumentation modules. The equipment allows fault-tolerant operation, improved fault detection, and self-diagnostics while the system is in operation.

Containment. The primary containment is a steel-lined reinforced concrete structure with a steel dome located within the reactor building. It is highlighted in black in Figure 3-5. Inside the primary containment is the pressure vessel, gravity-driven cooling system pool, suppression pool, and depressurization valves. The last three provide rapid response in the event of loss of coolant.

Modular Construction. Modularization techniques are proposed to reduce costs and shorten construction schedules to as little as 30 months. These

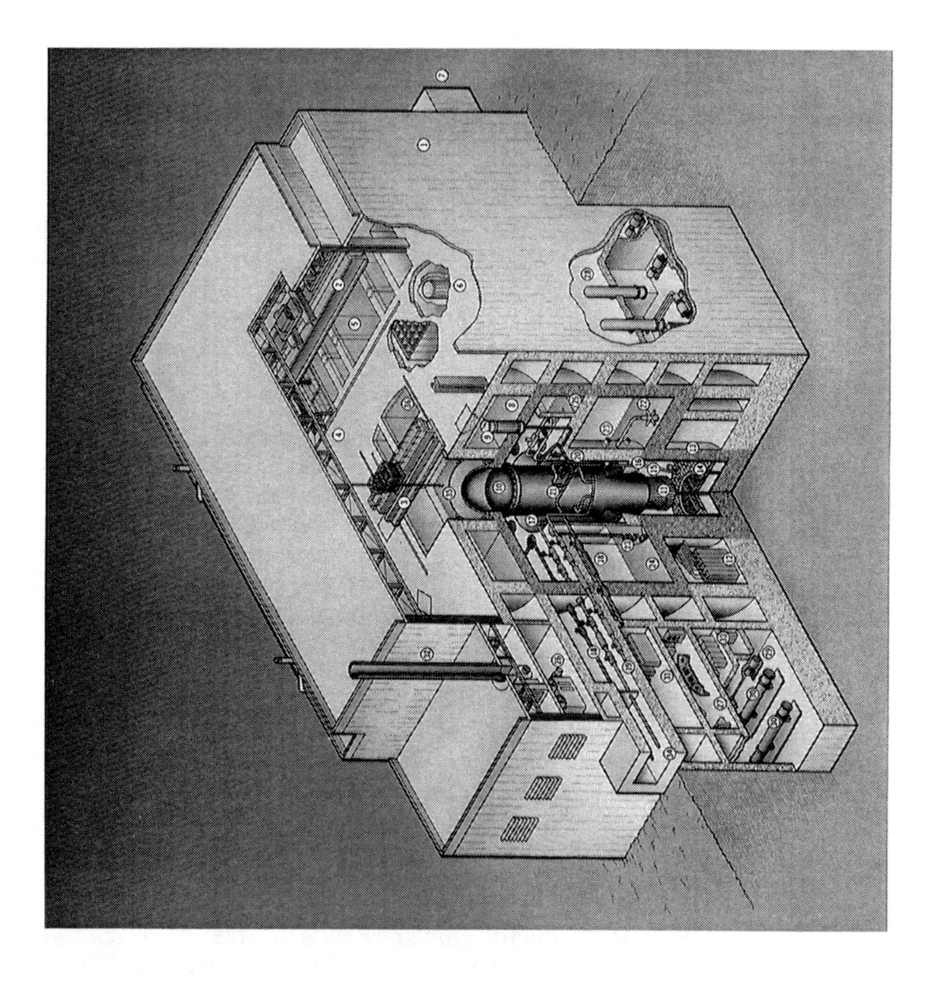

1 Reactor Building
2 Reactor Building Crane
3 Refueling Machine
4 Fuel Handling Machine
5 Spent Fuel Storage Pool
6 Spent Fuel Shipping Cask & Pool
7 Equipment Main Entry Hatch
8 Isolation Condenser Pool
9 Isolation Condenser
10 Reactor
11 Fine-Motion Control Rod Drives
12 FMCRD Hydraulic Units
13 Reactor Pedestal
14 Under-Vessel Servicing Platform
15 Lower Drywell
16 Shutdown Cooling Line
17 Upper Drywell
18 Main Steam Lines
19 Feedwater Lines
20 Depressurization Valves
21 Safety Relief Valves
22 SRV Quenchers
23 Horizontal Vents
24 Suppression Pool
25 Gravity-Driven Cooling Pool
26 Building HVAC
27 Control Room
28 Residual Heat Removal System Heat Exchangers
29 Reactor Component Cooling Water System Pump
30 Reactor Service Water System Heat Exchangers
31 DC Batteries
32 Plant Stack
33 FMCRD Electric Panel
34 Steam Tunnel
35 Drywell Head
36 Steam Separator Storage Pool

FIGURE 3-5 The simplified boiling water reactor. SOURCE: [GE Nuclear Energy, 1989]

techniques will be applied to reinforcing bar assemblies, structural steel assemblies, steel liners for the containment and associated water pools, and selected equipment assemblies, such as isolation condensers, drywell piping, heating, ventilation, and air-conditioning units, and water treatment equipment.

Other Reactor Concepts

CANDU-3 Heavy Water Reactor

HWRs are used in Canada for commercial electric power generation. These reactors are known as CANDU (for Canadian deuterium uranium) reactors. Although DOE operated HWRs for weapons material production for over 30 years, their design is very different from the CANDU design. For example, the current heavy water weapons material production reactor operates at room temperature with no significant pressure, and it has several annuli of fuel within a "universal sleeve housing." By contrast, the CANDU is a pressurized reactor, its fuel is within a "pressure tube," which itself is within a low pressure "calandria tube," and it operates at a high temperature.

The CANDU-3 is the latest and smallest version of the CANDU pressurized heavy water system developed in Canada.[AECL, Undated; AECL, 1989] Its steam supply system is shown in Figure 3-6. CANDU-3 has a net output of about 450 MWe and complements the established mid-sized CANDU 600 plant. A high level of standardization has been a feature of CANDU reactors. The vendor notes that, in CANDU-3, all key components, such as steam generators, coolant pumps, pressure tubes, and refueling machines, are identical to those in operating CANDU power stations. AECL states that the nuclear safety principles applied to the CANDU-3 reactor ensure that Canadian regulatory requirements are met. These requirements take the form of general criteria against which the developer must establish detailed design requirements.[AECL, Undated]

A letter of intent to submit the CANDU-3 design for standard design certification under 10 CFR Part 52 has been sent to NRC.

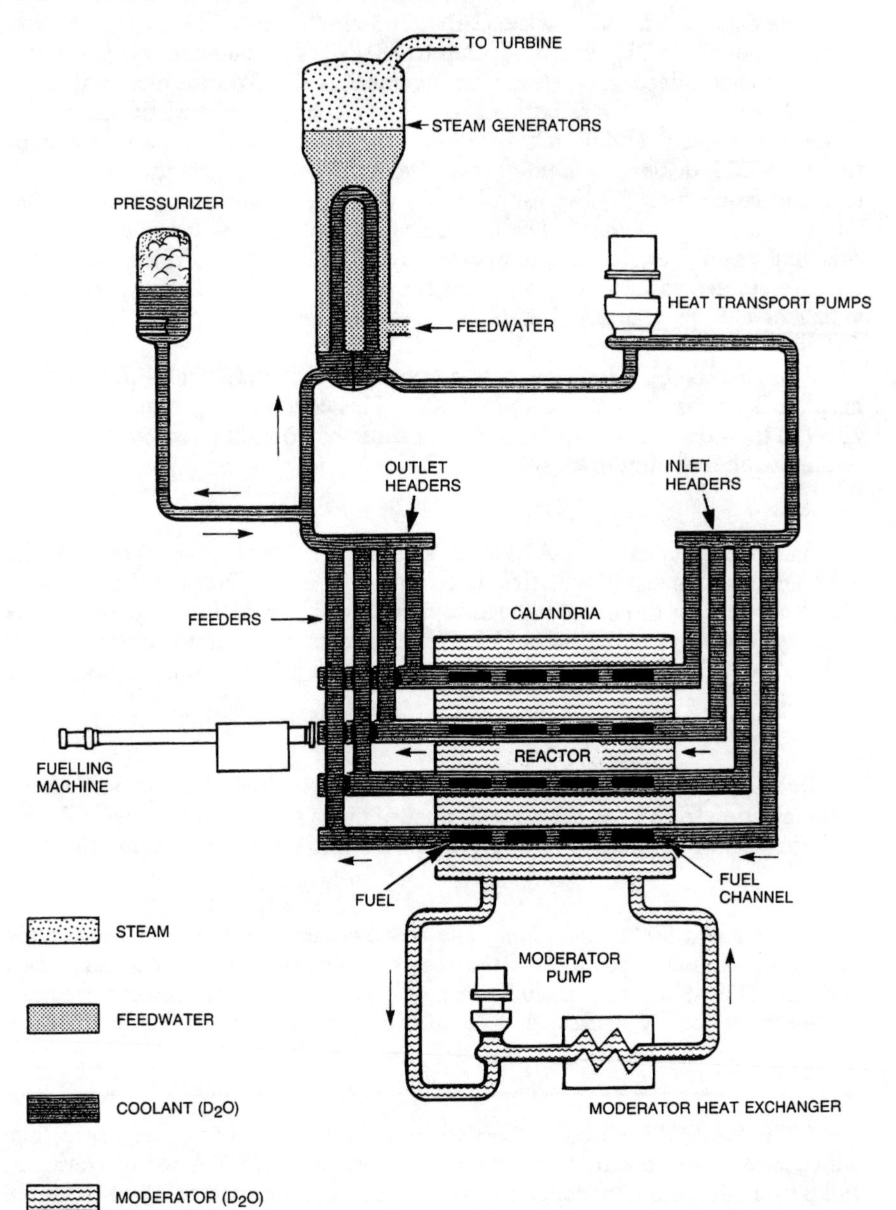

FIGURE 3-6 Steam supply system of CANDU-3. SOURCE: [AECL, Undated]

Core Design. Design of the CANDU-3 reactor core closely follows that of the larger CANDU reactors (of up to 881 MWe). The core design incorporates the standard geometrical arrangements of horizontal fuel-containing pressure tubes in a square lattice and has neutronic characteristics similar to those of current CANDU 600 reactors. There are three features unique to the CANDU designs, including the CANDU-3: (1) the use of natural uranium oxide fuel, (2) the use of heavy water as a moderator and coolant, and (3) on-power fueling.[2] The outlet header coolant pressure is about 1,450 psia and the outlet coolant temperature is about 590°F. These operating parameters are somewhat lower than the 2,250 psia and 615°F approximate values of U.S. pressurized LWRs.

The CANDU-3 design has a small positive void coefficient during a large break LOCA, as does the CANDU 600. This coefficient produces a power rise (50 to 100 percent per second) that must be counteracted by one of the two independent shutdown systems.

Steam Generators. CANDU-3 steam generators, like those of the CANDU 600, consist of a vertical U-tube bundle in a cylindrical shell, located above the reactor to ensure natural coolant circulation on loss of power to the primary cooling pumps. As in U.S. PWR systems, the heated coolant (heavy water in CANDU reactors) is contained on the primary side of the steam generator.

Fluid Systems. The ECCS operation includes provisions for both short-term injection from pressurized accumulator tanks and long-term recirculation of a mixture of ordinary and heavy water from the reactor building floor.

Control and Instrumentation. The reactivity control units are the reactor sensor and actuator portions of the reactor regulating and reactor shutdown systems. These systems include reactor power measuring devices, neutron absorbing reactivity control and shutdown devices, and the liquid injection

---

[2] The CANDU reactor has on-power (also known as "on-line") refueling, which means that the fuel is changed routinely with the reactor operating at full power. A fueling machine inserts new fuel into the reactor's fuel channels. A fuel transfer system brings new fuel into the reactor building and takes out irradiated fuel. Both the fueling machine and the fuel transfer system are automated and operated from the main control room. Surveillance equipment designed to monitor CANDU refueling operations is used by the International Atomic Energy Agency so that compliance with nuclear safeguards requirements can be verified.[AECL, Undated; AECL, 1989]

nozzles of the shutdown system. The shutdown system is physically and functionally separate from the regulating system.

All CANDU reactors use digital computers for the control of the reactor regulating system and other process systems, such as the pressurizer and steam boiler levels. In the CANDU-3, however, the two large central computers in the CANDU 600 systems have been replaced by a distributed control system consisting of a number of electronic modules distributed throughout the plant and linked by coaxial-cable data highways. This control system feeds data directly to color graphic operator stations, which form the interface between the operator and the plant. The design also features digital automated startup.

Containment. The CANDU-3 containment consists of a containment envelope of reinforced concrete with a full steel liner. All penetrations that are open to the atmosphere close automatically when an increase in containment pressure or radioactivity level is detected.

Modular Construction. The layout of a CANDU-3 power station permits modular construction because the contents of each building are subdivided into modules on a system and subsystem basis. The interfaces between modules are intended to facilitate site assembly and minimize site construction time. In addition, fuel channels can be factory assembled as can the steel calandria that contain the heavy water moderator. The shield tank, shield tank extension, and deck for the reactivity mechanisms are also amenable to off-site construction.

Safe Integral Reactor

CE has undertaken the design of the SIR jointly with Rolls Royce and Associates Limited, Stone and Webster Engineering Corporation, and the United Kingdom Atomic Energy Authority.[Bradbury et al., 1989] SIR is a PWR in which the reactor core, pressurizer, reactor coolant pumps, and steam generators are contained in a single reactor pressure vessel. The plant can produce a nominal station power output of 640 MWe from one turbine-generator supplied with steam from two identical 320 MWe pressurized LWR modules. Figure 3-7 illustrates the SIR design.

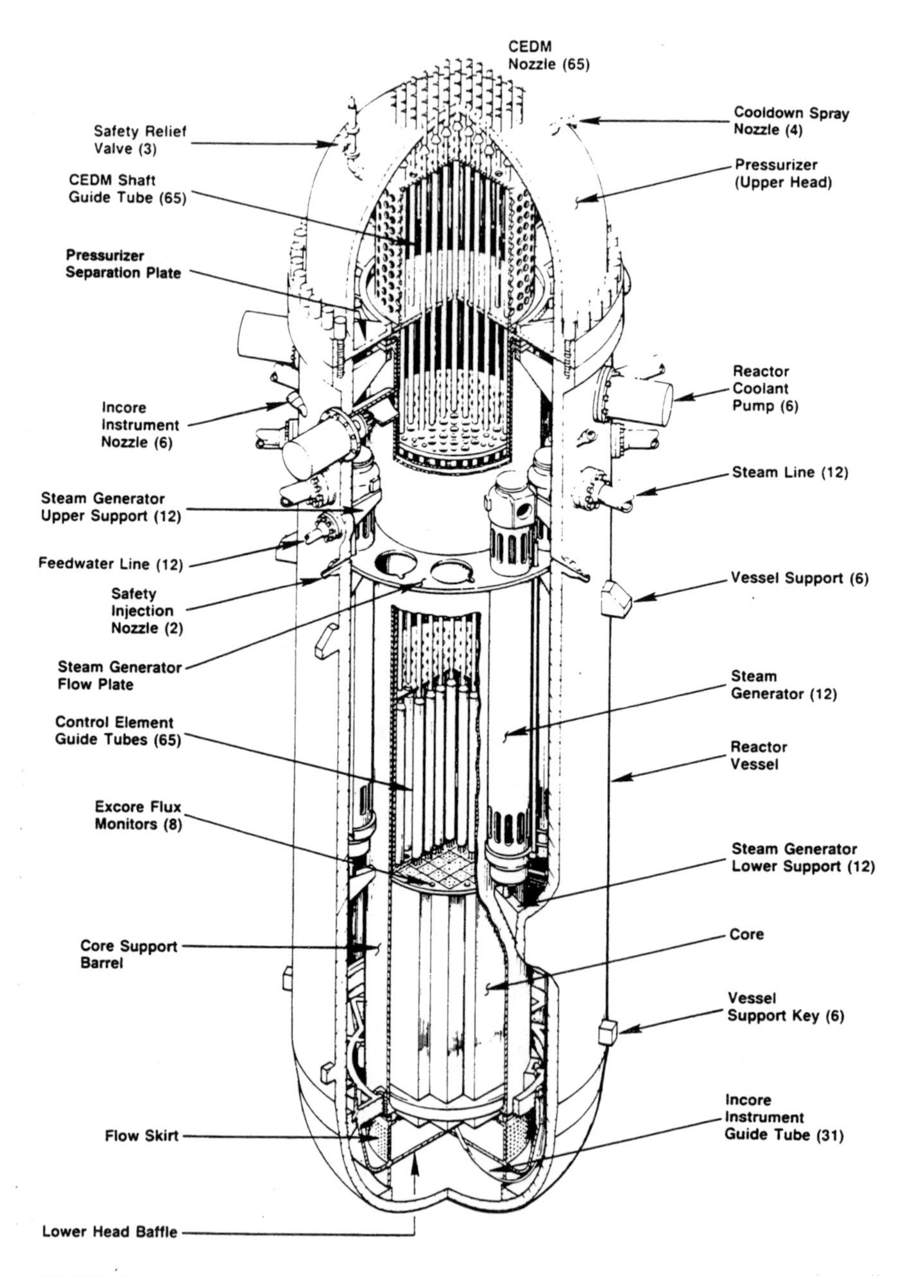

FIGURE 3-7a The safe integral reactor. SOURCE: [CE, 1989b]

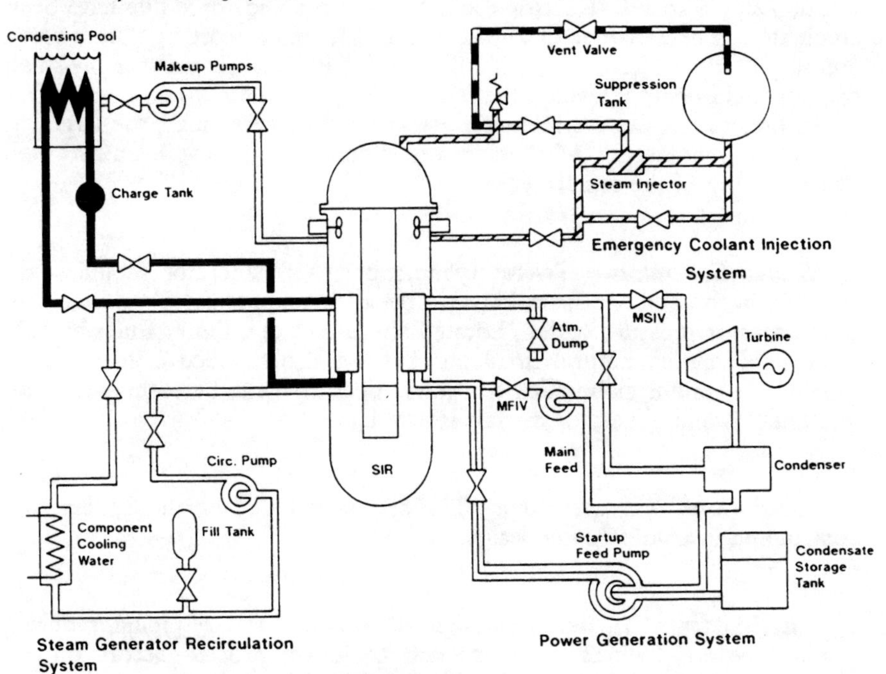

FIGURE 3-7b The safe integral reactor heat removal systems. SOURCE: [CE, 1989b]

Core Design. The reactor assembly is a completely self-contained PWR within a single vessel. Reactor coolant loop pipes and surge line have been eliminated. The SIR fuel, fuel assembly, in-core and ex-core instrumentation are all patterned after current CE designs for PWRs. By using many small components in parallel within the reactor vessel, primary system connections to the pressure vessel are relatively few and can be kept small; the largest is 2.8 inches in diameter. All pressure vessel penetrations have been kept well above the top of the reactor core.

Steam Generators. Twelve cylindrical steam generator modules are installed in the annular space between the core support barrel and the wall of the reactor pressure vessel. Located above the core, the modules provide the primary circuit natural circulation but are also shielded from the core. Finally, the vendor claims that full power operation can be maintained with one faulty steam generator module isolated.

Control and Instrumentation. This system is based on the System 80+ control and instrumentation design.

Fluid Systems. In the SIR design, there is no primary piping, reducing primary system failures. The normal cooldown process occurs on the secondary side, where subcooled fluid is circulated through the secondary side of the steam generators. For LOCAs, passive decay heat removal systems provide long-term cooling and are configured for a minimum of 72 hours of operation without intervention. The use of soluble boron for reactivity control has been eliminated.

Containment. The containment consists of (1) the reactor vessel compartment, which houses the reactor pressure vessel and support structure; (2) eight cylindrical steel pressure suppression tanks with external fins, each containing a pool of water; and (3) a vent system that connects the reactor vessel compartment to the pressure suppression tanks. The containment structure is filled with inert gas to prevent hydrogen ignition.

The reactor vessel compartment is a steel-lined, reinforced-concrete cylindrical structure capped by a removable steel dome. A vent pipe connects the gaseous space of the compartment to the shop-fabricated, cylindrical, steel pressure suppression tanks. These tanks are housed within a reinforced-concrete structure that has outside air intake and discharge ducts for circulating ambient air.

Modular Construction. The compact and simplified SIR design is suited to modular installation.[Bradbury et al., 1989] With the use of advanced construction techniques, the time from the first concrete pour to fuel loading is estimated to be 30 months.

Modular High-Temperature Gas-Cooled Reactor

In the United States the development of gas-cooled reactors has largely been the result of the efforts of DOE and a group of utilities supporting GA Technologies. The first helium-cooled reactor was a 40 MWe demonstration unit built at Peach Bottom, Pennsylvania, which operated from 1967 to 1974. A type of coated fuel particle was successfully used in this unit. A larger 330 MWe plant was built at Fort St. Vrain, Colorado. This unit was recently shut down because of steam header cracks, a low capacity factor due largely to poor circulator performance, and the resulting poor economics.

In Europe, development work in Germany has been led by Siemens and ABB. France and Great Britain were early pioneers in the use of reactors cooled with carbon dioxide. The German thorium high-temperature prototype reactor (THTR) produced approximately 3 billion kilowatt hours (kWh) of electricity. Some technical problems occurred during its operation (e.g., high friction of graphite balls). A reevaluation of continued operation of the THTR was made in late 1988. Considerations such as the termination of fuel supply, the inability to assure spent fuel storage, the possibility of additional requirements being imposed prior to obtaining a long-term license, and larger estimated decommissioning costs led the consortium that owns the plant to seek increased government participation or, absent that increase, to shut the reactor down. The THTR was shut down in late 1989.[3][Gas-Cooled Reactor Associates, 1989; Hill, 1989]

The advanced MHTGR concept of GA Technologies is a helium-cooled unit. The important features of the design presented to the Committee are

---

[3] Germany also operated a 15 MWe pebble-bed high-temperature gas reactor known as the AVR for about two decades. This reactor has been shut down.[Gas-Cooled Reactor Associates, 1989]

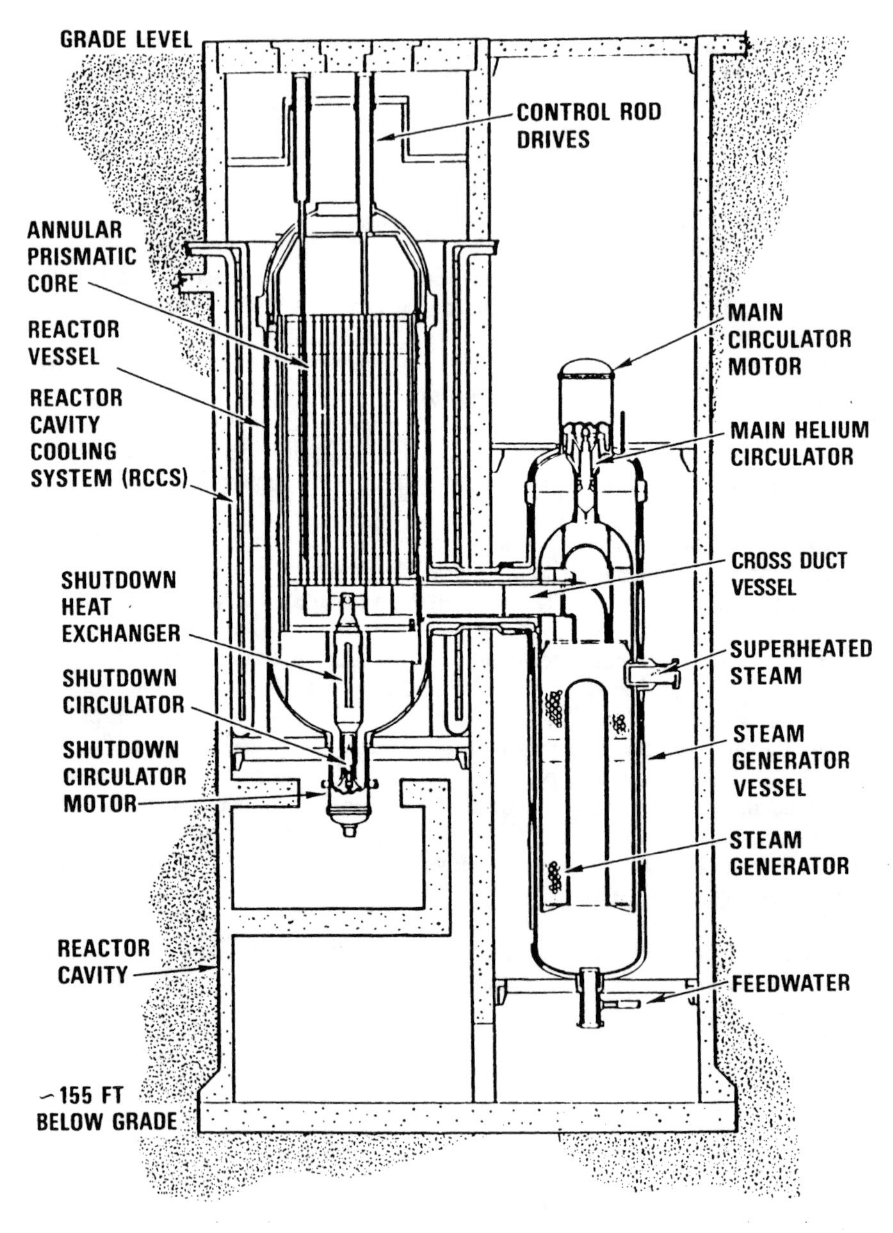

FIGURE 3-8 The advanced modular high-temperature gas-cooled reactor. SOURCE: [GA, 1989]

depicted in Figure 3-8.[4] This design generates 538 MWe from four nuclear modules and two turbine generators,[5] using steam at 2,515 psia and 1,005°F. The high temperature operation of the MHTGR leads to high thermal efficiency.[GA, 1989; Taylor, 1989; Nylan et al., 1988] An additional feature is its potential to provide process heat because of the high coolant exit temperature of 1,268°F. The current development is sponsored by DOE and Gas-Cooled Reactor Associates, with technical support from EPRI. This reactor's nuclear steam supply module is graphite-moderated and helium-cooled. The use of inert helium in contact with graphite core materials leads to low radioactive releases and low radiation exposure to workers in the plant if the helium coolant purity can be satisfactorily maintained. The conceptual design is presently under review by DOE for development as one of two reactor technologies for production of nuclear weapons materials as an eventual successor to the HWRs used at Savannah River.

Core Design. The reactor core is a low power density design that consists of an annular array of hexagonal blocks of graphite fuel elements surrounded by a reflector of unfueled graphite blocks. The design is intended to provide efficient heat transfer to the exterior in order to limit the temperature rise of the fuel in the event of a LOCA. The fuel consists of particles of uranium oxycarbide, enriched to about 20 percent in uranium-235, and thorium oxide. The fuel particles or kernels are about 0.8 millimeter in diameter, coated with porous graphite, and covered by successive layers of pyrolytic carbon, silicon carbide, and pyrolytic carbon. The coated particles are bonded together in fuel rods placed within sealed vertical holes in the graphite fuel element blocks.

The graphite fuel element blocks, together with the graphite moderator/reflector, provide a large heat sink in the event of an emergency. Preliminary data from temperature ramp tests of about 50°C per hour indicate essentially no failure of the refractory coating around the fuel particles below1800°C. Coating integrity at elevated temperatures for extended periods requires further evaluation.

---

[4] The Committee learned in mid-1991 that the MHTGR design has been changed. While the Committee did not have an opportunity to review the new MHTGR study, the Committee understands that the objective was to reduce costs while retaining the postulated safety advantages.[DOE, 1990] Thus, some of the design details listed below may no longer be current (e.g., a given module may produce more power). However, the Committee is not aware of any changes to the fundamental principles underlying the MHTGR concept discussed here.

[5] A possible new design could produce 692 MWe with four somewhat different nuclear modules and four turbine generators.[DOE, 1990]

LOCA simulation tests were conducted in the late 1980s on a small high-temperature gas-cooled reactor in Germany. The reactor was the experimental 15-MWe AVR (Arbeitsgemeinschaft Versuchs Reaktor). The most significant test demonstrated this reactor's safe response to conditions simulating an accident in which the coolant rapidly escapes from the reactor core and no emergency system is available to restore coolant flow.[Krüger and Cleveland, 1989]

Steam Generator. A single steam generator per reactor module is located in a separate steel vessel. The once-through shell and tube design uses helically wound tubes to carry water in at the bottom and steam out at the top. After passing through superheater sections at the top, the steam is discharged through a nozzle assembly in the upper side wall of the steam generator.[Gas-Cooled Reactor Associates, 1987] Although previous work on the use of a helium turbine in a closed cycle to eliminate the steam generating system was abandoned about ten years ago, developments in high-temperature gas turbines have prompted renewed interest in this concept.

Fluid Systems. If the active cooling system is inoperable in an emergency, decay heat can be dissipated by conduction and radiation to the reactor cavity cooling system in the reactor enclosure. This system circulates atmospheric air by gravity to ultimately remove the decay heat. If the reactor cavity cooling system defaults, passive radiation and conduction transport heat directly to the silo structure and surrounding earth.

Control and Instrumentation. Plant control is based on a fully integrated system in which one operator monitors automatic startup, operation, and shutdown of the two-unit power module. Such a distributed control system would employ the latest technology in computer and communication technology, and system operating procedures would reside in software on local process controllers while overall plant performance was governed by a supervisory computer.[Gas-Cooled Reactor Associates, 1987; EPRI, 1989a]

Containment. The reactor, as presently configured, is located below ground level and does not have a conventional containment. The absence of a containment for the proposed commercial reactor is a major issue, especially given that DOE plans to have a containment for the proposed new production reactor version of the MHTGR.[6][Beckjord, 1989] The basic rationale of the designers is that a containment is not needed because of the safety features inherent in the properties of the fuel that were discussed previously. Regarding the possibility of including a containment building for the commercial version, DOE stated to NRC:

> Because of its enhanced safety characteristics, the MHTGR has such a high level of safety that no further meaningful improvement in public risk can be obtained at reasonable cost.[Williams et al., 1989]

NRC has not yet made a determination of the acceptability of the proposed MHTGR design without a containment.

Modular Construction. It is claimed that each of the four reactor modules can be factory-fabricated. GA Technologies and Bechtel estimate that

---

[6] According to DOE, "The primary reason that the MHTGR-NPR [new production reactor] containment system is different from the commercial MHTGR, is to avoid dependence of the development of the NPR on successful completion of the technology program that is necessary to validate assumptions made in the commercial program. The NPR is developing the design and supporting technology in parallel coordinated efforts. These efforts require a decision on the containment system prior to completion of technology efforts that would substantiate the commercial reactor containment approach. The commercial program does not have this constraint.

In addition there are significant differences between the commercial MHTGR and the NPR that justify different design selections to meet requirements. These differences include:

- Provide the NPR-MHTGR with additional flexibility to accommodate unforeseen future missions.
- Accommodate the different reactor core which utilizes highly enriched fuel without thorium, includes production materials, and has a different operating cycle."[Young, 1989]

DOE also stated, "The development of the commercial MHTGR is prepared to be stretched out if there are technology development delays associated with validating the plant design without a low leakage containment structure. Verification of the performance characteristics of high quality fuel is a significant element of the justification for not requiring a low leakage containment structure for the commercial MHTGR."[Young, 1989]

construction could be completed in four years from the issuance of a construction permit.[Taylor, 1989]

Process Inherent Ultimate Safety Reactor

The PIUS reactor, a 640 MWe pressurized LWR development, is sponsored by ABB Atom and originated in Sweden. Stated safety features of PIUS are (1) safety ensured by the laws of mechanics and gravity, (2) lack of actively actuated components, (3) lack of required operator action, (4) insensitivity to human errors and malicious intervention, and (5) ability to withstand violent external events.[Bredolt et al., 1988] Figure 3-9 illustrates the PIUS design.[ABB, 1989]

Core Design. PIUS, in the early stages of design, is a passive PWR immersed in a large prestressed concrete pressure vessel filled with cool, borated water at about 1,340 psia.

The reactor is contained in a cylindrical structure that extends from the bottom of the core, near the bottom of the vessel, to the top enclosure. During normal operation this structure separates the circulating hot coolant loop from the cool vessel water by two hydraulic density locks.[7] The coolant loop has a low concentration of boron, in contrast to the vessel water. During normal operation, the heat generated in the reactor is carried by the coolant upward to the top of the cylinder and then to a steam generator, where the main coolant pump returns it on a flow path inside the cylindrical structure to a point below the core. The reactor power is controlled by the temperature and the boron content of the reactor circulating loop. There are no control rods in the PIUS 600. If the main coolant pump stops, the water circulates by natural circulation through the density locks, bringing the cool borated water into the core and shutting down the reactor.

The fuel assemblies are standard pressurized reactor fuel elements with low-enriched uranium oxide pellets in fuel rods.

---

[7] A hydraulic density lock makes use of the principle that water separates naturally into layers that have different densities. The application of that principle in PIUS means that during operation cold borated water sits below the core while lower-boron content hot water in the primary loop passes over the cold water and through the reactor. Loss of circulation in the primary loop results in the cold, highly borated water being drawn into the core through the chimney effect, thus shutting down the reactor.

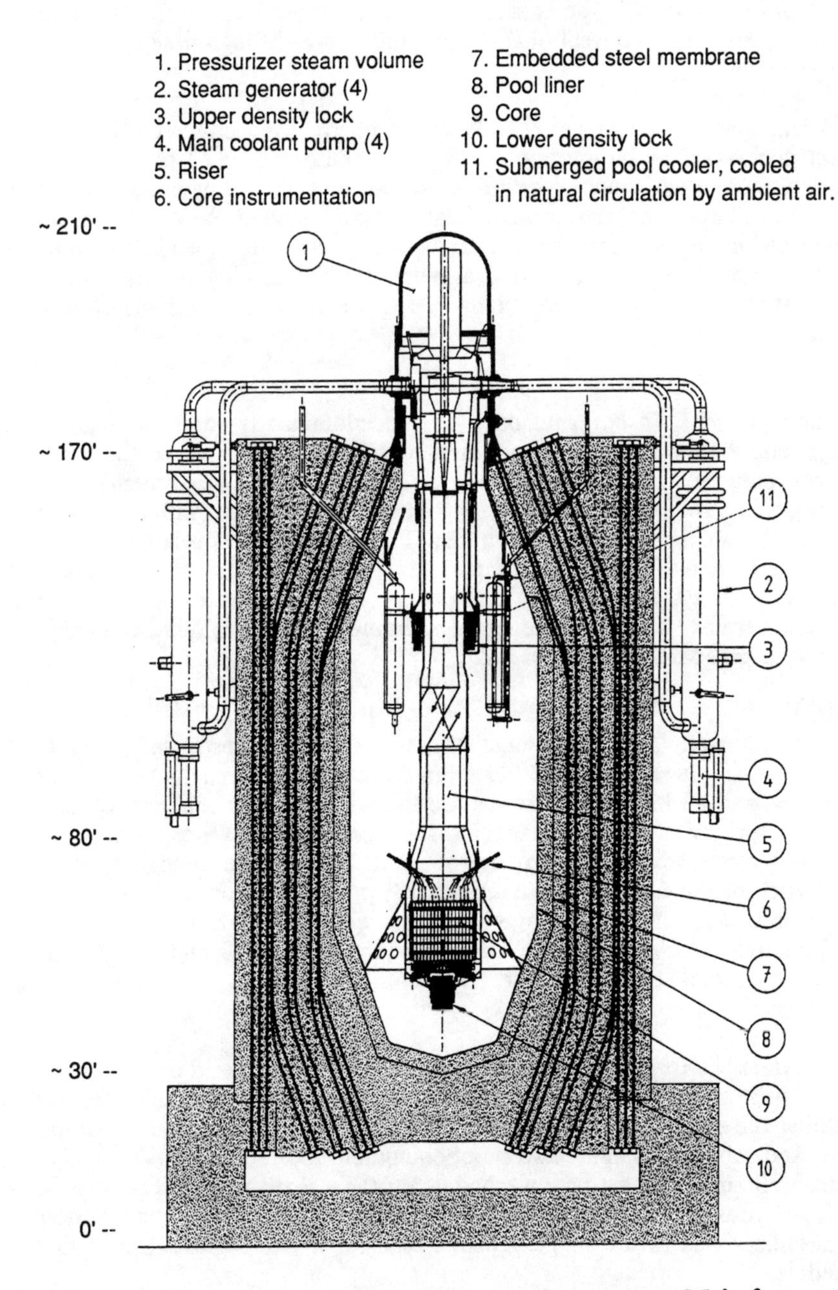

FIGURE 3-9 The process inherent ultimate safety reactor. Main features of the nuclear steam supply system. SOURCE: [ABB, 1989]

Steam Generators. The steam generators, located outside the concrete reactor vessel, use a conventional straight-tube once-through design.

Fluid Systems. Residual core heat can be removed either by the four steam generators or by the pool. Pool heat removal is used for extended shutdowns or in emergency conditions and can be achieved by either passive or active means. In the passive heat removal system, heat exchangers submerged in the pool transfer heat to the secondary side, which is cooled by naturally circulating ambient air drawn from a dry cooling tower. Water temperature can be maintained below 100°C under these conditions even in the case of a large LOCA.

Control and Instrumentation. The predominantly non-safety-grade equipment, based on micro- and mini-computers, is located in the main control room area and is distributed in the plant (decentralized system). The safety-grade parts (e.g., the reactor trip and interlock system with associated measuring systems and control equipment for initiating safety-related actions) are located in two separated compartments at the bottom of the reactor building. All systems are implemented on microcomputers, arranged in redundant trains. Man-machine interactions are based on color video display units with keyboard and tracker balls.

Containment. The containment structure is a large, prestressed-concrete reactor vessel in which the cold borated water, the reactor core/riser assembly, and all key safety systems are located. The key characteristics of this structure are that it (1) contains sufficient borated water to cool the reactor for one week after reactor shutdown, (2) is large enough to store spent fuel for the lifetime of the reactor, (3) provides a high level of protection against saboteurs, (4) contains both steel reinforcing bars and prestressed steel tendons, which together provide a very strong structure, and (5) contains a double internal steel liner to prevent water leakage.

Modular Construction. The construction of PIUS is based on separation of building units, prefabrication of parts of the containment and pressure vessel at the site, use of conventional process systems, limited use of pumps, pipes and cables, and limited scope of equipment located inside the containment. A 36-month construction schedule for the *n'th* plant is predicted based on BWR construction experience and does not rely on modular construction. A modularization review suggests possibilities for shortening the construction schedule.

PRISM Liquid Metal Reactor

Fast reactors normally use liquid sodium as a coolant. They can produce more fissile material than they consume and are often referred to as "breeder" reactors. LMRs have been used to produce electricity in the United States, France, Great Britain, the Soviet Union, and Japan.[Collier and Hewitt, 1987]

In the United States, a small experimental breeder reactor (EBR-I) built by Argonne National Laboratory generated the first electricity from nuclear fission in 1951 and in 1953 confirmed that breeding was possible. In 1955 the second core of this reactor was partially melted during an experiment designed to investigate its prompt positive reactivity feedback coefficient. Fuel rod bowing was determined to be the cause, and subsequent core designs corrected that problem. Subsequently, a second experimental breeder reactor (EBR-II) was built by Argonne National Laboratory in Idaho and began operation in 1963. It has demonstrated the practicality of the LMR design in which the entire primary system is submerged in a pool of sodium. Since the mid-1960s the EBR-II has been a test facility for LMR fuel assemblies and structural material irradiation and safety tests.

In April, 1986 two significant safety tests were conducted at EBR-II. These involved loss of flow without scram from full power and loss of heat sink without scram from full power. These tests successfully demonstrated the safety potential of the integral fast reactor (IFR), a generic reactor technology defined by the use of liquid sodium as coolant and metallic uranium and plutonium as fuel. The reasons for the safe responses illustrated in the EBR-II tests are inherent to the IFR. Specifically, the properties of the metallic fuel and the large thermal inertia of the sodium pool are key to achieving reactor shutdown passively (i.e., without relying on operator intervention, active components such as control rods, pumps, valves, or the use of balance of plant for heat removal) while keeping temperatures low.[Chang, 1989]

While the capability to ride out a loss of flow without scram from full power and a loss of heat sink without scram from full power add markedly to the safety of an LMR, the presence of a positive sodium void coefficient in the present design has been a matter of concern. This has led to the addition of rod stops to control reactivity insertions that may result in sodium boiling.

If an advanced LMR is proposed having a significant positive sodium void reactivity worth, careful evaluation will have to be made of the effect of this attribute on the severity of postulated accidents involving reactivity insertion or other events which could lead to sodium boiling. Additional or redundant features may be necessary to remove this concern.

The Committee understands that additional or alternative features to remove this concern are being identified. The positive void coefficient of the currently proposed LMR design [PRISM] results from a design criterion for a shippable reactor vessel that in turn determines the maximum core diameter and the required core height for the specified power module. To eliminate the possibility of the positive void reactivity worth, revisions to the reference PRISM design have been suggested that would result in a larger diameter core with a lower height. Although this arrangement would require a field-fabricated vessel, the elimination of the undesirable positive void coefficient characteristic may be deemed worthy of the loss of a shop-fabricated, rail-shippable reactor vessel.

The more recent fast flux test facility (FFTF) constructed at the Hanford site is a loop-type LMR. It has been used to test full-length oxide fuel assemblies and for limited tests on advanced metallic fuel assemblies. The first commercial LMR built in the United States, Fermi-1, was also a loop-type reactor. It suffered melting of two fuel assemblies in 1966 as a result of a flow-channel blockage. Although Fermi-1 was repaired and restored to operation, it was eventually decommissioned.[Collier and Hewitt, 1987]

The French liquid metal program has constructed three reactors of increasing size culminating in the commercial size SUPER PHENIX plant (SUPER PHENIX was built by a consortium of several European countries). The first two plants, RAPSODIE and PHENIX, performed well and provided valuable experience upon which to build the French program. However, the RAPSODIE experimental reactor was shutdown subsequent to discovery of a tiny leak on the primary sodium circuit, the repair of which was considered too expensive to justify maintaining the reactor in service after 15 years of operation. The 250 MWe demo-plant PHENIX, which started regular operation in 1974, is shut down pending study and evaluation of transient negative reactivity pulses observed in 1989 and 1990 while the reactor was operating at full power. Operation of the SUPER PHENIX 1,200 MWe plant has been curtailed by a sodium leak, discovered in 1987, in an auxiliary vessel for the storage of discharged fuel. This facility is being replaced, and the new one should be ready by the end of 1991.

PRISM is a modular, passively stable, advanced LMR being designed by GE. Its present design uses a new metal alloy fuel being developed concurrently by Argonne National Laboratory as part of the IFR program. The IFR concept includes the first reactor, fuel reprocessing, and fuel fabrication using reprocessed fuel. PRISM is a specific design of generic IFR technology. [Berglund, 1989; Till, 1989] The nuclear steam supply system for a PRISM reactor module is depicted in Figure 3-10.

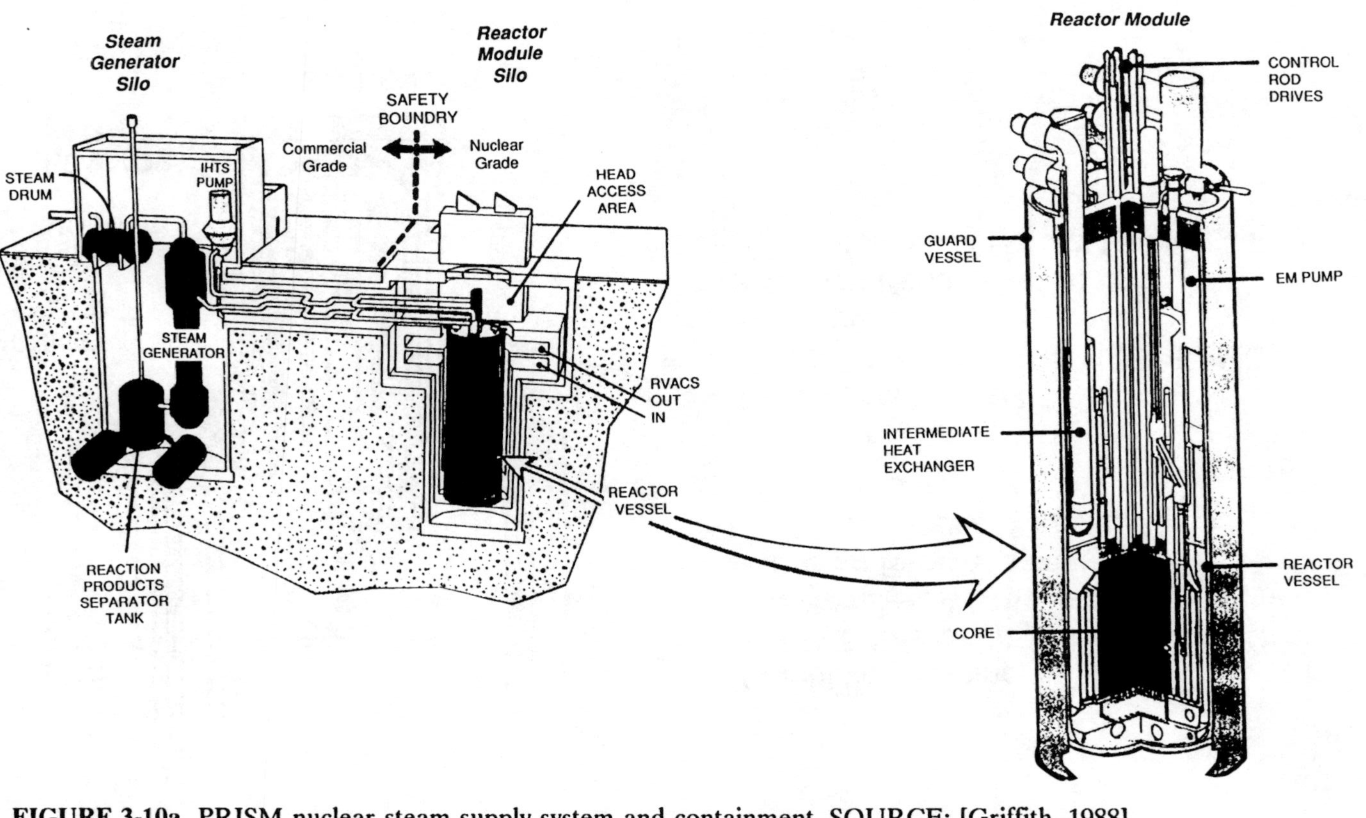

FIGURE 3-10a PRISM nuclear steam supply system and containment. SOURCE: [Griffith, 1988]

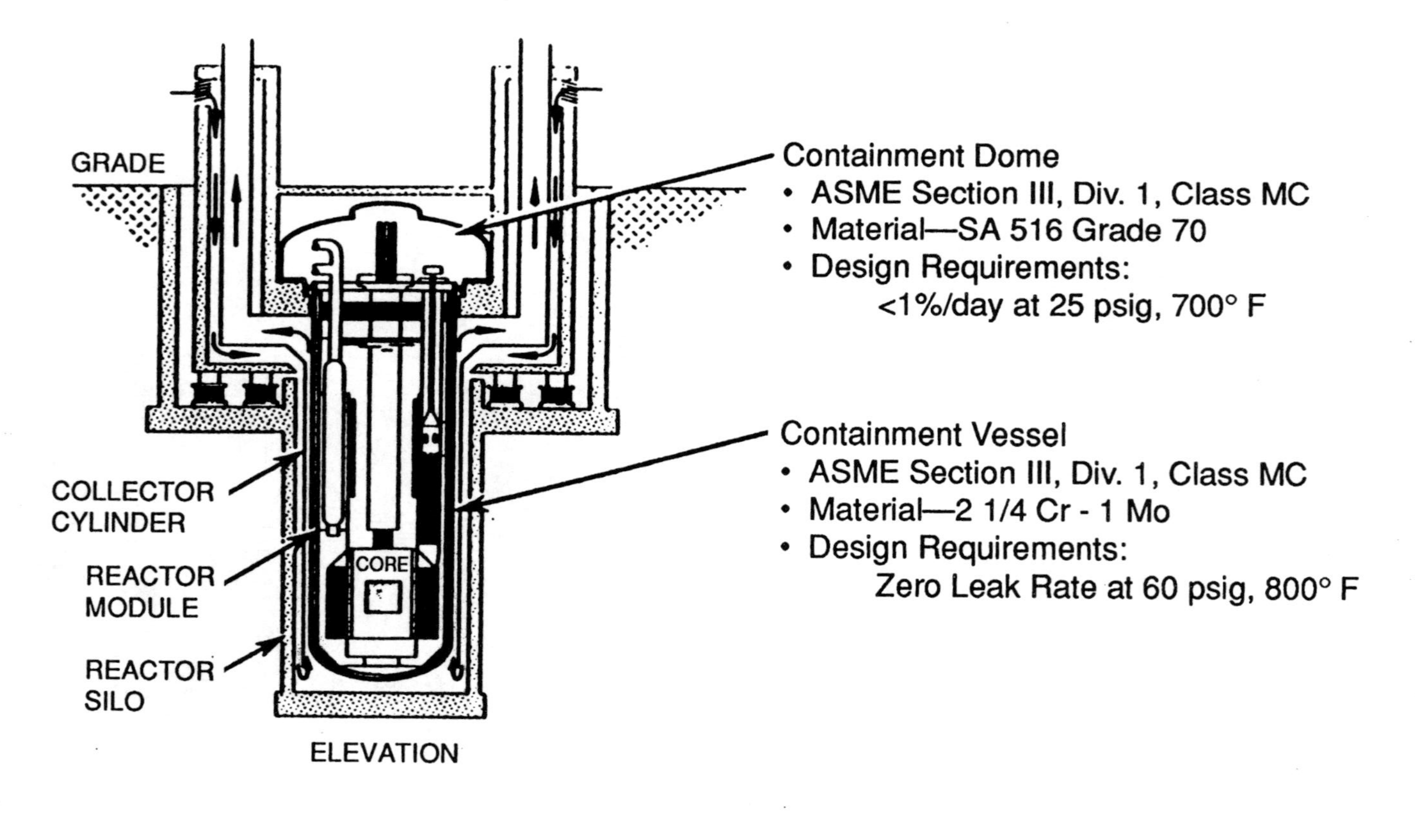

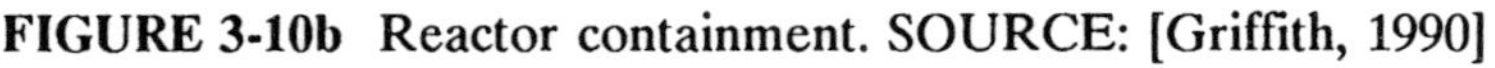

FIGURE 3-10b Reactor containment. SOURCE: [Griffith, 1990]

Core Design. The PRISM reactor plant is made up of one to three 465 MWe power blocks, each with three 155 MWe reactor modules. (The plant design in *ALMR Design and Program Summary* and *The Liquid Metal Reactor* [Berglund, 1989; Till, 1989] uses three power blocks.) The reactor core for each module is in a pool of liquid sodium, which is circulated through the core by four cartridge-type electromagnetic pumps. The pool system consists of a large tank of sodium into which the reactor core, sodium pumps, and two intermediate heat exchangers are placed. The tank is in a guard vessel, which would collect sodium if it were to leak from the pool. This feature assures that the core will remain covered and cooled by sodium.[Berglund, 1989; Till, 1989]

The heat from the reactor module is transferred from the primary sodium coolant loop to a secondary sodium loop through two intermediate heat exchangers. In this way radioactive sodium in the primary loop is isolated from the steam generator. The sodium in the secondary loop enters a single steam generator that produces steam for the turbine generator. The three steam generators for a power block feed steam to a single 465 MWe turbine generator.

The PRISM reference fuel is a uranium-plutonium-zirconium alloy with plutonium concentrations of about 25 percent. As discussed earlier in connection with the safety tests at EBR-II, the properties of metallic fuel are a major contributor to the passive safety features of the PRISM design. Argonne National Laboratory is also developing a pyrometallurgical reprocessing system, in connection with the IFR concept, which could lead to fuel reprocessing and recycling.

Steam Generator. A single-wall helical coil steam generator is believed to provide high reliability (less than one failure per sixty year plant life for a nine unit plant) and economic operation.[Nuclear Power Assembly and ANS, 1990] The steam generator system provides early warning of a tube leak, and an isolation and pressure relief system to limit the sodium-water reaction damage from such a leak, or from multiple tube leaks.

Fluid Systems. The reactor vessel auxiliary cooling system provides emergency core cooling after any incident that impairs the normal emergency heat removal systems. This auxiliary cooling system removes residual heat by radiant heat transfer from the reactor pool to the guard vessel to atmospheric air, which is always circulating upward around the guard vessel. Passive reactor stability is inherent because of a large negative temperature coefficient of reactivity. This combination of passive cooling and passive reactor stability ensures residual heat removal without operator intervention. Thus, if all cooling through the intermediate heat exchangers is lost and the control rods do not automatically shut down the reactor, the negative temperature coefficient of reactivity will bring the reactor to an equilibrium state at a low power level where heat removal from passive systems will maintain the fuel temperature low enough to prevent fuel damage.

Control and Instrumentation. The plant control system, which is not safety grade, provides a high level of automation for normal plant operation and utilizes redundant digital equipment and power supplies to operate nine nuclear steam supplies, three turbine generators, and associated plant equipment from a single control center. Startup, operation, and shutdown of each module are automated. There is a safety grade reactor protection system for each reactor that performs all safety grade functions, including scramming, and is independent of and isolated from the control system.

Containment. The PRISM is located under ground-level. Based on the latest information provided by DOE, the PRISM advanced LMR design includes a lower containment vessel and an upper containment dome (see Figure 3-10b). The lower containment is intended to contain reactor pool leaks, while the upper dome is intended to mitigate severe events postulated to cause an expulsion of radionuclides into the region above the reactor. The dome is made of steel that is 1 to 1-1/2 inches thick, and the lower containment consists of 1 inch thick steel.[Griffith, 1990]

Modular Construction. The design includes compact reactor modules sized to enable factory fabrication, economical shipment to both inland and water-side sites, and full-scale prototype testing. Balance of plant modules contain structures and equipment, piping, electrical wiring, and related components.

In Situ Metallurgical Reprocessing of Fuel. Following removal of test fuel pins from the core, an electrorefining process extracts a uranium-plutonium mixture, including fission products producing high dose rates, from the dissolved mixture of fuel, steel, and fission products at temperatures around 550°C. The blanket material is electrorefined in such a way that uranium alone is processed to enrich the product in plutonium. The processed blanket material can then be added to the electrorefined fuel, which is always radioactive.

Actinide Transmutation. Actinides, or the elements in the series beginning with actinium (89) and ending with lawrencium (103), include several very long-lived radioactive alpha emitters and are among the materials of greatest concern in nuclear waste disposal beyond 300 years, depending on the site characteristics and the scenario assumptions under consideration.[Till, 1989;

NRC, 1991a] The actinides, in a nuclear reactor or possibly an accelerator designed for the purpose, can be transmuted, with the production of fission energy, to radionuclides that often have a much shorter half-life. An advanced LMR, having no moderator in the core and hence a faster neutron energy spectrum, has much more favorable actinide cross sections than a thermal reactor. An LMR can recycle its own actinides and also actinides from LWR spent fuel, operating as an actinide burner or a breeder, if desired. [Till, 1989; Chang et al., 1987] (A thermal reactor is more limited in the extent to which it can transmute actinides.)

The Committee notes that there exist previous studies of hazards and risks from radioactive waste disposal which have found that, for a given site and a given set of assumptions about repository characteristics and the severity of natural and man-made events, technetium, not the actinides, introduces the greatest risk in the long term.[National Research Council, 1983 and 1984] Since some additional technetium would be the result of recycling the actinides, the net effect would be the production of energy and a proportionate amount of additional technetium, which would still have to be placed in a repository so as to provide long-term safety.

Figure 3-11 is an overview of DOE's proposed actinide recycling process. No such processes for LWR spent fuel recycle have been demonstrated to date, although recycling of plutonium-uranium (mixed oxide) fuel has been demonstrated. Substantial further analysis and research is required to establish (1) whether high-recovery recycling of transuranics and their transmutation can, in fact, benefit waste disposal, and (2) the technical and the economic feasibility of recycling in LMRs actinides recovered from LWR spent fuel.[8][Pigford, 1990] The Committee notes that a study of separations technology and transmutation systems was initiated in 1991 by the DOE through the National Research Council's Board on Radioactive Waste Management.

---

[8] In late 1990 Professor Thomas Pigford distributed a paper on actinide burning and waste disposal that raised many questions about the technical and economic aspects of recycling actinides in liquid metal reactors.[Pigford, 1990] The Committee has not performed a technical review of that paper but believes that Pigford's analysis supports the need for a careful and objective evaluation of whether the development of transuranic recycle and transmutation, if successful, will actually benefit the geologic repository. Pigford's analysis should be considered carefully by those advocating actinide recycling as a solution to the high-level waste disposal problem. The Committee notes that DOE has provided comments disagreeing with aspects of the Pigford paper.[Young, Undated]

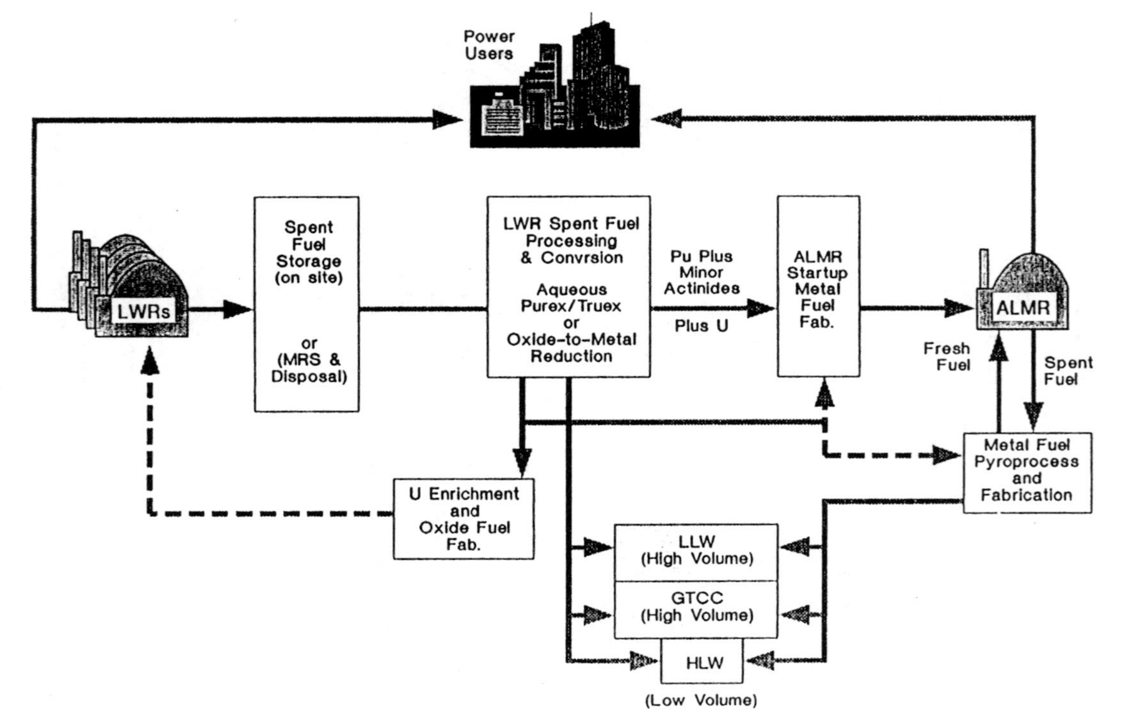

FIGURE 3-11 Advanced liquid metal reactor (ALMR) actinide transmutation recycling of light water reactor fuel. SOURCE: Actinide Recycle, Presentation to National Research Council Committee on Future Nuclear Power Development, Office of Nuclear Energy, U.S. Department of Energy, January 29, 1990

EVALUATION OF THE TECHNOLOGIES

The Committee developed the following criteria for comparing the advanced reactor technologies and furnished them to presenters before they briefed the Committee:

- safety in operation;
- economy of construction and operation;
- suitability for future deployment in the U.S. market;
- fuel cycle and environmental considerations;
- safeguards for resistance to diversion and sabotage;
- technology risk and development schedule; and
- amenability to efficient and predictable licensing.

More detail on the criteria is provided in Appendix B. Vendor estimates related to the criteria are presented in Table 3-3.

The Committee believes that the broad criteria listed above represent the considerations that are (a) most able to be influenced by a choice of technology, and (b) significant to a future determination of whether or not one or more of the advanced reactor technologies is deployed in the United States. For example, the discussion in Chapter 2 has established clearly that the safety and economics of nuclear power substantially affect its acceptance by the public, government, and the private sector.[9]

The Committee's evaluation was performed by assessing all of the technologies with respect to each broad criterion, starting with safety. (It should be noted that not all subordinate entries in Appendix B were explicitly addressed by the Committee, either because of a lack of specific data or because they were judged to be of lesser importance to the choice of reactor technologies.) The results of the evaluation follow. A summary appears at the end of each section, and the entire evaluation concludes with an overall assessment. The information available with which to perform evaluations is uncertain and often promotional, as should be expected for designs that exist

---

[9] In the context of studying energy research and development strategies for reducing emissions of greenhouse gases, a National Research Council Committee has suggested that future reactors should be subject to a set of international criteria developed from an international study "on criteria for globally acceptable reactors."[National Research Council, 1990] Illustrative issues for which criteria would be established include safety, reliability, scale, simplicity and standardization, waste disposal and storage, diversion resistance, cost, and fuel efficiency (i.e., issues similar to those considered in this report).

TABLE 3-3 Estimates Provided by Vendors Related to Evaluation Criteria [a]

Reactor	Core Damage Frequency [j]	Availability Goal (percent)	Plant Life (years)	Overnight Capital Costs (1989 $ per rated kWe) [g]	Levelized Generating Costs (1989 cents per kWh) [k]	Construction Time (first concrete to on line, in months) [n]	Projected Date of NRC Certification	Projected Date of Lead Plant Operation [b]
Large Evolutionary Light Water Reactors								
ABWR	$<10^{-6}$	86	60	940-1,190	3.0-3.3	48	1991[e]	1996 (in Japan)
APWR-1300	$<10^{-5}$	90	40	1,350	3.9	54	1995	2000
System 80+ PWR	$6x10^{-7}$	87	60	<1,143	3.7	54	1992[e]	Late 1990s (in UK, Korea, or U.S.)
Mid-Sized Light Water Reactors With Passive Safety Features								
AP-600 PWR	10^{-6} to $4x10^{-6}$	90	60	1,300	3.9	42	1994[e]	2000
SBWR	Goal $<10^{-5}$	90	60	1,200-1,500	3.5-4.1	36	1995[e]	2000
Other Reactor Concepts								
CANDU 3 HWR	Goal $<10^{-6}$	94	100[c]	1,713	5.5[d]	30	1993 [p]	1996 (In Canada)
SIR PWR	$6x10^{-7}$	87	60	<1,475	<4.3	36	1998	Late 1990s (in UK)
MHTGR[m]	(note i)	80	Not specified	2,000[f]	5.2[f]	44	2002	1998 (Demonstration plant)
PIUS PWR	(note h)	90	60	1,400	4.0	42	Uncertain [p]	Uncertain
PRISM LMR	To be determined	90	60	1,000-1,375	3.1-5.1	31	2003	1998 (Prototype)

[a] For comparison with EPRI requirements, see Table 3-2.
[b] Unless otherwise specified, plant location would be in the United States.
[c] Modular component life is 40 years.
[d] Figures provided by AECL were revised to exclude spent fuel management costs, consistent with other vendor estimates.
[e] The ABWR and System 80+ projections are likely to slip 3 years, and the AP-600 and SBWR projections are likely to slip to 1996, according to NRC staff.[NRC, 1991b]
[f] These figures are 1988 dollars for an nth of a kind plant.
[g] Overnight costs exclude time related costs such as interest.
[h] Vendor has found no credible incidents leading to core degradation.
[i] PRA implicitly states no fuel-failure sequences down to 10^{-8} per year. Confirmation of this low probability of fuel failure must be sought in a future PRA based on more detailed design information.[Williams, 1989]
[j] Does not include external initiators such as earthquake and flood.
[k] Variable costs over several decades (30 to 40 years) are levelized, i.e., converted to equivalent constant annual cost over the same time period. Costs exclude spent fuel management.
[m] Information provided to the Committee in 1989. A report [DOE, 1990] on a more recent design was subsequently provided. This report, based on the vendor's estimates, indicates improved economics.
[n] Assumes no licensing delays.
[p] Late-1990 NRC projections for PIUS indicate 1997 as a possibility; CANDU appears to be extended indefinitely past NRC's earlier projection of late-1996 because its prototype is deferred.[NRC, 1990b; NRC, 1990c]

SOURCES: Vendor presentations to Committee and follow-up communications. The Committee has not evaluated the accuracy of these estimates.

principally in concept or with non-prototype testbeds. Thus, the Committee concluded that numerical rankings would give a false sense of accuracy. Consequently, the overall assessment represents the Committee's qualitative judgments as a result of considering all the criteria together. The criteria were used primarily in two ways: (1) to provide an outline of issues for the vendors to use in developing their presentations and submissions to the Committee, and (2) to provide a framework for the Committee to discuss the alternative technologies. The Committee concluded that it would not be appropriate to provide weightings for each criterion and then to grade the approaches, add up the scores, and get a selection.

Safety

Discussion

About three-quarters of the nuclear power plants in operation worldwide are situated outside the United States, and this fraction is growing. Another accident anywhere will have major negative consequences for the development of nuclear power worldwide. International cooperation on safety among utilities, suppliers, research organizations, and licensing authorities is therefore necessary.[10] The Committee notes that the International Atomic Energy Agengy (IAEA) has established means to monitor the safety performance of nuclear power plants, including the classification of safety significant events (International Nuclear Event Scale).

In the design of future advanced LWRs, vendors are guided by the safety design policy presented in the Requirements Document prepared for EPRI. The safety design policy states that "there will be excellence in safety both to protect the general public and to assure personnel safety and plant investment protection."[EPRI, 1990] While the safety record of existing nuclear power plants has been very good, more ambitious safety targets have been established for future advanced LWRs. (See Table 3-2 for a summary of EPRI's advanced LWR design requirements.) Safety in the advanced LWR program extends well beyond hardware-oriented lessons learned from existing plants. Attention is focused on areas such as plant simplification, design margins, human factors, and an integrated approach to safety.[EPRI, 1990]

Each reactor designer presented current but only partial design information to the Committee. Where available, probabilistic risk assessments (PRA) were preliminary and did not benefit from detailed system design. Until full

---

[10] Some believe that next generation nuclear plants will be international efforts subject to international safety standards.[Chung and Hazelrigg, 1989]

PRAs of actual reactors (including external events) are available and subjected to careful and extensive peer review, there will not be a satisfactory basis to compare the relative safety of the different concepts. Furthermore, the absence of detailed engineering design and the lack of construction and operating experience with the actual reactor concepts make a meaningful, quantitative safety comparison less achievable. In particular, PRA is not a sufficient basis to compare the safety of new concepts with that of proven concepts due to the lack of reliability data of active and passive components sufficiently based on experience. However, if final safety designs of advanced reactors, and especially those with passive safety features, are as indicated to this Committee, an attractive feature of them should be the significant reduction in system complexity and corresponding improvement in operability. While difficult to quantify, the benefit of improvements in the operator's ability to monitor the plant and respond to system degradations may well equal or exceed that of other proposed safety improvements.

The Committee believes that each of the concepts considered can be designed and operated to meet or closely approach the safety objectives currently proposed for future, advanced LWRs, albeit with the considerable uncertainty inherent in risk assessment and in estimates for this extremely low projected level of risk.[Lewis, 1978; NRC, 1990a] If design goals are realized, these plants will be safer than existing reactors. The different advanced reactor designs employ different mixes of active and passive safety features to achieve the safety objectives, and there is, of course, more experience with certain designs than others. The Committee believes that there currently is no single optimal approach to improved safety. There is a distinct advantage to passive containment cooling for preventing containment failure due to slow over-pressurization. However, dependence on passive safety features does not, of itself, ensure greater safety, especially given the potential effects of earthquakes, design errors, inspectability, manufacturing defects, and other subtle failure modes. Consequently, the Committee believes that a prudent design course retains the historical defense-in-depth approach.

In most future reactors, defense-in-depth would be achieved by a multiplicity of safety barriers and features, including a containment structure to mitigate the consequences of core damage accidents. However, one advanced reactor type (the MHTGR), without a containment structure, was proposed. The Committee was not convinced by the presentations or the material supplied to support them that the core damage frequency has been demonstrated to be low enough to make a containment structure unnecessary.[11]

---

[11] The Committee notes that, at present, the new production reactor (NPR)-MHTGR program includes a containment. If the MHTGR is selected for the NPR, containment-accident scenario analyses will proceed more

Summary

The Committee could not make any meaningful quantitative comparison of the relative safety of the various advanced reactor designs. All of the designs are claimed to achieve safety that equals or exceeds the levels specified by EPRI in Table 3-2 (e.g., $<10^{-5}$/year core damage frequency). If design goals are realized, these plants will be safer than existing reactors. Dependence on passive safety features does not, of itself, ensure greater safety; the historical defense-in-depth approach must be retained. In particular, for the MHTGR, the Committee was not convinced that a containment structure is unnecessary.

Economy

Discussion

Vendor-estimated overnight capital costs and levelized generating costs are shown in Table 3-3 for the reactor technologies that the Committee examined. Most of the estimates for generating costs are based on a 30-year levelized cost analysis, including capital carrying charges, fuel, and operations and maintenance (O&M) (see definitions in Chapter 2). The uncertainties in overnight capital costs and levelized generating costs are quite large because different cost models and assumptions were used for their calculations. Also, U.S. experience with LWRs provides little assurance that construction of the large evolutionary reactors will meet cost and schedule claims.

Vendor estimated overnight capital costs (in dollars per kilowatt electric) and levelized generating costs (in cents per kilowatt hour) for CANDU are higher than those estimated for all LWRs. The higher estimated costs for the CANDU reactor may be partly related to the use of a different cost model than that used by other vendors. Another factor is that CANDUs have been built--the CANDU-3 is quite similar--so AECL has real data to use, unlike some of the other vendors. Additionally, the designs of all the advanced (except possibly for the evolutionary) reactors are still in the stage where cost estimates change. In particular, SIR, MHTGR, PIUS, and PRISM have a very high degree of economic uncertainty. For the different types of evolutionary reactors, levelized generating costs and overnight capital costs are likely to be similar.

---

rapidly.

EPRI has independently evaluated some overnight capital costs and O&M costs.[EPRI, 1989b] The estimates are more general than those of the vendors, but they are based on clear definitions. Uncertainties are estimated as -30 to +80 percent. The results in Table 3-4 show that, except for the MHTGR, EPRI's estimates of overnight capital costs are somewhat higher than those of vendors shown in Table 3-3.

The large evolutionary LWRs have higher estimated total construction costs and longer construction times than the mid-sized LWRs with passive safety features, but, as shown in Tables 3-3 and 3-4, they are estimated to be competitive on a cost per kilowatt electric basis. Estimated construction times under ideal conditions for the mid-sized LWRs with passive safety features ranged from 36 to 42 months (Table 3-3), but there are serious uncertainties about meeting the claimed construction schedules, which in turn could have a major impact on the total funding required to complete the plant. Even though some utilities may prefer to order the larger plants, the perceived larger financial risk may be a deterrent to their deployment.

To reduce construction and operating costs, designers of the advanced mid-sized plants have attempted to simplify their designs, adopt modularized construction, and reduce construction times. However, because there is no experience in building such plants, cost projections for the first plant[12] are clearly uncertain. To reduce the economic uncertainties it will be necessary to demonstrate the construction technology and improved operating performance.

Some mid-sized LWRs currently in operation have demonstrated consistently high capacity factors.[IAEA, 1990] Consequently, estimates that assume advanced versions of the same size can also achieve high capacity factors may prove to be correct. (Table 3-3 shows availability projections in the range 80 to 94 percent for all advanced designs and about 90 percent for the mid-sized LWRs with passive safety features. Availability is usually within a few percent of capacity factor. Availability and capacity factor are defined in Chapter 2.) Because the newer heavy water CANDU reactors are a refinement of currently operating reactors, their claimed capacity factors should be attainable. However, capacity factors of the other reactor concepts, SIR, MHTGR, PIUS, and PRISM, have a very high degree of uncertainty.

[12] The descriptive term "first plant" refers to a plant that will be demonstrating new technological features in design, construction, or operations. It is potentially the first commercial operating reactor of this design and, as such, has performance uncertainties in construction and operation. It represents a commercial technology demonstration.

TABLE 3-4 EPRI-Estimated Overnight Capital and Operations and Maintenance Costs (In December 1988 Dollars)

Advanced Reactor Type	Overnight Capital Costs (per rated kWe)	Operations and Maintenance Costs	
		Fixed[a] ($/kWe-yr)	Incremental[b] (cents/kWh)
Large evolutionary light water reactors	$1,300	61.1	0.11
Mid-sized passive light water reactors	$1,475	72.7	0.11
Liquid metal and high-temperature gas-cooled reactors	$1,725	75.5	0.15

[a] These operating costs are essentially independent of actual capacity factor, number of hours of operation, or amount of kilowatts produced. They include labor charges for plant staff.
[b] These variable operating costs and consumables are directly proportional to the amount of kilowatts produced. They include chemicals consumed during plant operation.

SOURCE: EPRI. 1989. Technical Assessment Guide, Electricity Supply–1989. EPRI P-6587-L, Volume 1: Rev. 6, Special Report, September.

The MHTGRs are estimated to have higher capital costs than the other plants, and they may have higher operating costs, as shown in Tables 3-3 and 3-4.[13] Moreover, if NRC were to mandate a conventional containment, that requirement could adversely affect the economics of this reactor design as well as the technical feasibility of its passive cooling feature.

LMR plants (e.g., PRISM) may be able to compete economically with water reactors if fuel reprocessing (being developed as part of the integral fast reactor program) turns out to be technically and economically feasible, and if the overnight capital costs of these plants are as low as the vendor indicates. (For the IFR, reprocessing would be *in situ* pyrometallurgical, but for the LMR concept in general, reprocessing options include centralized plants as well as aqueous technology.)[Nuclear Power Assembly and ANS, 1990] EPRI cost estimates (Table 3-4) suggest that these capital costs will be higher.

---

[13] Recent design changes intended to reduce costs of the MHTGR while retaining its postulated safety advantages imply that the MHTGR economics may be more favorable than reported here. The MHTGR Cost Reduction Study Report [DOE, 1990] states that the modified MHTGR could be cost-competitive with the AP-600. The Committee has not analyzed such projections, nor has EPRI produced a review of them, but notes that they substantiate the large uncertainty in economic projections for advanced reactors. Furthermore, the Committee assumes all reactor designers are working on improvements of the designs and concepts presented to the Committee.

Neither the PRISM design nor the PRISM technology are sufficiently developed to provide a reasonable degree of confidence in cost estimates. Finally, different institutional arrangements may be required for utility involvement in a PRISM plant because of reprocessing, concerns about diversion of sensitive nuclear materials, and lack of utility experience with the technology.

Summary

The economic projections are highly uncertain, first, because past experience suggests higher costs, longer construction times, and lower availabilities than projected and, second, because of different assumptions and levels of maturity among the designs. The EPRI data, which the Committee believed to be more reliable than that of the vendors, indicate that the large evolutionary LWRs are likely to be the least costly to build and operate on a cost per kilowatt electric or kilowatt hour basis, while the high-temperature gas-cooled reactors and LMRs are likely to be the most expensive. EPRI puts the mid-sized LWRs with passive safety features between the two extremes.

Market Suitability

Discussion

None of the reactor concepts the Committee reviewed is likely to be operating in the United States before the year 2000. If the large evolutionary LWRs being built in Japan (and perhaps in Korea) perform well, market potential in the United States will be improved. Large U.S. utilities with several nuclear power plants are likely to be the first customers for such plants if they need large base load electrical generators and if financial risks are acceptable.[14]

Compared to the large evolutionary reactors, the mid-sized advanced pressurized and simplified boiling water reactors with passive safety features have lower total overnight capital costs (but not lower costs per kilowatt electric), hence less total capital at risk, but no construction and operating experience. The smaller size of these plants might be attractive to a larger number of possible purchasers.

---

[14] Some Committee members believe that the large evolutionary LWRs will be the next nuclear plants to be ordered in the United States, because of perceived economies of scale and greater confidence by utilities and investors in making modest extensions of proven technology.

The heavy water CANDU reactor has been marketed in Canada and other countries. The main barriers to CANDU's competitiveness in the United States are the uncertainty of its licensing by NRC and the inexperience of U.S. utilities with heavy water technology. On the other hand, the earlier CANDU reactors have a good performance record and could be attractive to certain power producers, particularly if Atomic Energy of Canada, Limited were an investor.[AECL, Undated] It is difficult to weigh all these factors, but the Committee judges that this technology ranks below the advanced mid-sized LWRs in market potential.

The Committee believes there is no near-term U.S. market for the other LWR concepts, SIR and PIUS. While SIR is based on proven light water technology, there are serious uncertainties about the operations, maintenance, economics, and possibly safety of a system configuration that is substantially different from that of current plants. Also, the SIR design appears less complete than the AP-600 or SBWR. The level of testing or prototyping that would be required by NRC is unclear. The PIUS reactor is viewed as a preliminary design with no relevant experience. It is the Committee's view that experience with other LWRs is not relevant to PIUS. While there is no regulatory experience related to PIUS, a conceptual design for this reactor was submitted to NRC for an informal licenseability review. The lack of operational and regulatory experience for both SIR and PIUS is expected to significantly delay their acceptance by utilities, especially if positive experience has been obtained for the evolutionary large reactors or mid-sized LWRs with passive safety features.

The market potential of the MHTGR is very difficult to evaluate. Although gas-cooled reactors have been available for more than 20 years, they have not had commercial success. The strategic advantage of the MHTGR is its high temperature, which permits high temperature process heat applications. However, siting requirements and the extent of a U.S. market for that capability are unclear. The overnight capital cost of the modular design is relatively high on a per kilowatt basis (Tables 3-3 and 3-4). Further, considerable research and development (R&D) is still required for this advanced reactor, particularly on fuel pellet integrity and on reliable components, and a first plant for demonstration would be required. The issue of whether the design would require a containment building is still not resolved. If no containment building were needed, and the emergency planning zone was reduced to the site boundary, the MHTGR could have significant siting advantages that would make it more competitive. However, based on the Committee's view on containment requirements, and the economics and technology identified above, the market potential for the MHTGR was judged to be low.

Finally, the LMR might be commercially competitive if uranium fuel shortages limit the use of LWRs. The LMR's safety features and ability to recycle actinides are not considered important positive factors for its early market potential. Any strategy requiring fuel reprocessing introduces significant technical, economic, and non-proliferation policy considerations, some of which would complicate licensing.

Summary

The evolutionary LWRs and mid-sized LWRs with passive safety features are judged to have the highest market potential in the United States, while CANDU has the next highest. The other LWR concepts (SIR and PIUS) and the MHTGR are judged to have low U.S. market potential. Finally, the unique properties of the LMR might lead to a U.S. market, but only in the long term.

Fuel Cycle

Discussion

Fuel cycle evaluation encompassed three issues: (1) use of enriched fuel versus use of natural uranium as a fuel, (2) disposal of high-level radioactive waste, and (3) whether fuel reprocessing is needed.[15] The environmental implications of the technology derive, in large part, from these fuel cycle issues. Enrichment is important at the front end of the fuel cycle, and the disposal of high-level waste is important at the back end. Reprocessing can influence both the back end (waste disposal) and the front end (need for new uranium fuel). The Committee considered these issues to have roughly equal priority. Again, only enhanced and novel features of advanced reactor designs are discussed.

All LWRs, including SIR and PIUS, have essentially the same fuel cycle and corresponding environmental implications. Reprocessing of spent fuel to

---

[15] Reprocessing is not now considered economical in the United States for any reactor technology. Whether it will be so in the future is uncertain. The LMR is the only design that is presently considered for deployment as a breeder, in which event reprocessing would, of course, be necessary. If reprocessing is needed, technical, economic, and non-proliferation issues will have to be resolved.

recover enriched uranium or plutonium is not currently planned. None of these designs provides a substantially higher burnup than the others.

The CANDU design presented to the Committee uses natural uranium and does not require fuel enrichment; therefore, CANDU does not produce the low-level wastes associated with uranium enrichment. However, it has lower burnup, so the volume of spent fuel rods to be stored will be greater than in the case of the LWRs. In other aspects of fuel cycle management, the heavy water CANDU is comparable to the LWR.

The MHTGR presented to the Committee was designed to use fuel enriched with uranium 235 to about 20 percent versus only a few percent for the LWRs. The fuel pellets provide encapsulation of the waste, which might represent an additional barrier to release of the fission products. However, data to support this have not yet been obtained, nor has a strategy or process for the unique features of MHTGR waste disposal yet been developed. Reprocessing is not currently planned but may become necessary or desirable. At this time, there is no experience with reprocessing of this type fuel, although preliminary development of reprocessing requirements has been investigated. On balance, there does not appear to be a significant fuel cycle advantage or disadvantage to this reactor design.

Finally, the proposed LMR fuel cycle has the potential for substantial economic gains compared to LWR fuels. If a shortage of uranium develops the reactor could breed plutonium.[16] However, the feasibility of using this reactor as a breeder in a reprocessing-recycling manner requires policy, technical, and economic development and evaluation. A range of issues needs to be addressed in such a study, including LWR reprocessing as a source of additional fuel and the economics of LWR and MHTGR designs with high conversion ratios.[17] Assuming success, it would still be necessary to dispose of high-level waste, although the waste would consist of fission products, most of which, except for technetium, carbon, and some others of little import, have half-lives very much shorter than the actinides.

---

[16] The LMR breeder could be fueled from stockpiled depleted uranium once it has been started with plutonium or enriched U-235 from some external source. This would remove environmental problems associated with mining uranium and managing the associated mill tailings.

[17] As noted earlier, DOE has initiated a study of separations technology and transmutation systems.

Summary

Although there are definite differences in the fuel cycle characteristics of the advanced reactors, fuel cycle considerations did not offer much in the way of discrimination. All LWRs were judged about equal. Compared to the LWRs, CANDUs and MHTGRs had disadvantages at one end of the fuel cycle, but possible advantages at the other. The LMRs offer advantages because of their potential ability to provide a long-term energy supply through a nearly complete use of uranium resources.

Safeguards and Physical Security

Discussion

Safeguards regarding nuclear material in reactors and other facilities must be considered against diversion of fissionable material to nuclear weapons purposes, against sabotage of the power and reprocessing plants leading to a serious accident and release of radioactivity, and against terrorist theft and use of highly radioactive material as a terror weapon.

The problem of diversion is usually considered most serious when the facilities are located in countries that have a motivation for developing nuclear weapons. IAEA has developed an international safeguards regime, including on-site inspections and permanent inspection equipment. The IAEA system is applied to nuclear material at all sites in those non-nuclear weapons states that are party to the Nuclear Non-Proliferation Treaty. It is an obligation of the states to inform IAEA of the relevant sites. This application to all material at all sites is called full-scope safeguards. Many suppliers, including the United States, require such safeguards for exports to any non-weapon states. Other countries, including the nuclear weapon states, have safeguards applied to some, but not all, facilities. It is encouraging that Brazil and Argentina have recently agreed to safeguards on all facilities. The most important constraints for limiting proliferation of nuclear weapons are the political will of non-weapon states to forego weapons-development, the safeguards on nuclear (fissile) materials, and agreements by nations possessing advanced technology not to transfer nuclear weapon-related equipment or knowledge to non-weapon states. However, although such supplier agreements can limit the export of technologies that can be used to develop nuclear weapons, theft of weapons-grade material remains a threat. Accordingly, physical security must be provided for nuclear material, especially when in a form (i.e., enriched uranium or reprocessed plutonium) that is suitable or can readily be made suitable for weapons purposes. Physical security is also vital when nuclear material in storage or transit is susceptible to theft and use for terrorist purposes.

No country with nuclear weapons, or suspected of having nuclear weapons, has developed these weapons using fissionable material from a civilian power reactor, although civilian power programs have been used as a cover for other activities aimed at developing nuclear weapons.[18] (Dual-use reactors, producing both nuclear-weapons material and electricity, have been used however.) Nevertheless, reactors designed and employed for the production of power remain of concern for proliferation because they can be used for production of weapons grade plutonium. Some power reactors have even been designed to operate with highly enriched uranium or with plutonium as an initially-loaded fuel. Therefore, any fuel cycle must be examined for the possibility of diversion of weapons-grade material, or of material that could be further processed to produce weapons-grade material. In particular, the existence of centrifuge or laser enrichment techniques may make the path to weapons much easier, especially since almost all countries have access to natural uranium. In the future, technologies developed to permit efficient extraction of specific isotopes of plutonium may also facilitate the extraction of that element from spent fuel removed from reactors. This would create additional paths to diversion. In addition, deployment of any new fuel cycle in the United States or any other nuclear weapon state should be examined with a view to avoiding poor precedents in terms of proliferation. Fuel cycles should be designed to minimize diversion opportunities and maximize safeguardability, regardless of the country in which they are implemented.

The once-through fuel cycle where low enrichment fuels are used and the whole fuel rods, together with radioactive fission products, are buried, has the lowest potential for diversion of sensitive nuclear materials. The use of reprocessing[19] where plutonium is separated from the radioactive fission products makes the plutonium easier to use, although it must be noted that with normal burnup the presence of Pu_{240} in the plutonium makes the plutonium much more difficult to use in a reliable bomb. The opportunities for diversion are greater in any concept where on-line fuel loading is possible (i.e., the CANDU). Because it permits an operation where the fuel is removed more easily from the core after only a short time, and without as much Pu_{240} being built up, the consequences may also be greater.

---

[18] The Committee recognizes that one or more countries may have used, or may be using, plutonium from power reactors for the production of nuclear weapons. However, the Committee knows of no case where the weapons were initially "developed" using such materials.

[19] Reprocessing could, in theory, be used with any of the reactor concepts under consideration. However, it is required only if the LMR is deployed as a breeder.

While the high-temperature gas-cooled reactor is envisioned to use higher enriched fuel than the LWRs, the enrichment levels are below weapons-grade. Nevertheless, because much less separations work is needed to convert such material to weapons-grade, its use could increase proliferation concerns.[20] The MHTGR's proposed once-through fuel cycle would make the diversion risk not much greater than for the LWRs. In the LMR based on the integral fast reactor concept, reprocessing would take place in a closed system in which fuel containing actinides is produced. This fuel is highly radioactive and requires special handling in such a way that diversion would be difficult. However, the large amount of plutonium in use may require special safeguards, particularly if *in situ* reprocessing were not used. In particular, LMRs fueled by plutonium would pose a serious safeguards question in countries of proliferation concern. As compared to LWRs, the CANDU reactor poses some additional risks of diversion because of two features: (1) replacing fuel while the reactor is running increases access to fissionable material, especially plutonium, and (2) production and transportation of heavy water provides access to material that is useful in producing weapons-grade material.

Sabotage is always a threat against an industrial facility, posing a risk to the workers and, for some facilities, to the neighboring public. Power reactors pose a hazard because of the large fission product inventory, once the reactor has run for any significant length of time. Sabotage is another way of defeating safety systems. In order to prevent a knowledgeable person, and particularly a knowledgeable group of persons, from causing serious damage to a nuclear power plant by shutting off critical pumps and/or destroying safety systems, appropriate physical security measures must be and are taken. The advanced mid-sized reactors, even with their passive features, do not eliminate the problem of sabotage, but more detailed evaluation of the risks is needed. The other new LWR concepts and MHTGR appear to be even more resistant to a sabotage-induced fission product release. The LMR also has natural barriers to damage from sabotage, but in the reprocessing cycle some significant damage could be done. The concept of "defense-in-depth" that the Committee endorses for other reasons also provides barriers against acts of sabotage, although special design measures are possible that would further reduce the likelihood of successful sabotage without degrading safety.

---

[20] In raising natural uranium, which contains 0.7 percent U235, to an enriched state containing 93 percent U235, approximately 90 percent of the separative work is expended in reaching an enriched state containing 20 percent U235. Thus, only 10 percent more work is needed to reach the 93 percent U235 state.[American Physical Society, 1978]

Summary

The problems of proliferation and physical security posed by the various technologies are different and require continued attention. Safeguards and physical security considerations do not offer much discrimination among the reactor technologies, particularly for deployment in the United States. However, the CANDU (with on-line refueling and heavy water) and the LMR (with reprocessing) will require special attention to safeguards. Reactor designs that have passive safety features, including the high-temperature gas-cooled reactor and LMR, provide additional protection against certain acts of sabotage. However, special attention will be necessary to ensure that the LMR's reprocessing facilities are not vulnerable to sabotage or to theft of plutonium.

Development Risks

Discussion

The large evolutionary LWRs offer the most mature technology and are in various stages of design certification as standardized plants by NRC. The use of mature technology has many advantages:

- prior experience provides a check on estimates of reliability, maintainability, safety, and economics;
- construction and operations experience may not be required before NRC certification as a standardized design or before electricity producers would place an order for the design;
- technically qualified and skilled personnel are currently available;
- a related infrastructure exists for design, component manufacture, construction, and operation that is directly transferable;
- to begin operation, required procedures and training would not be drastically different from those used for the most recent LWRs; and
- almost all current regulations and regulatory experience relate to this type of reactor, so new federal funding is unlikely to be required to complete the certification process, but could accelerate the process.

For mid-sized pressurized and simplified boiling LWRs with passive safety features, successful design certification by NRC may depend in part on the outcome of contractual work recently initiated by DOE. However, regardless of government assistance, some research and much development and design are still required for these reactors. The research and design to demonstrate the passive safety features must be completed before certification. While these reactors are based on many years of LWR experience, they differ from current reactors in construction approach, plant configuration, and safety features. These differences do not appear so great as to require that a first plant be built for NRC certification.

The extent to which additional demonstration will be required by NRC for any new design has not yet been determined. However, in 1991 the Commission stated:

> The Commission approves in principle the requirement for prototype testing of new, innovative technology such as the nuclear power plant control room designs intended for design certification, if the testing is required to confirm expected operational performance under normal and abnormal conditions and thus is essential for the [NRC] staff's safety determination.[Chilk, 1991]

While a prototype in the traditional sense probably will not be required, federal funding will likely be required for the first mid-sized LWR plant with passive safety features. The level of government assistance required to build such a first plant is uncertain but could be significant. The Committee believes that the designs of mid-sized LWRs can be certified by NRC without construction of a prototype plant. However, federal funding is likely to be required to assist in the construction of the first mid-sized LWR plant; such funding would serve to offset some of the factors associated with the innovative features of these designs, such as the risks of not meeting the shortened construction times, the costs of first-of-a-kind engineering, and the uncertainties in the NRC licensing process.

The CANDU-3 reactor is farther along in design than the mid-sized LWRs with passive safety features. However, it has not entered the NRC design certification process. Commission requirements are complex and different from those in Canada so that U.S. certification could be a lengthy process.[Ahearne, 1989] Of particular note is the small positive void coefficient during a LOCA. NRC has always required strong negative void coefficients.

Development risks for the other LWR concepts (SIR and PIUS) are greater than those for the technologies discussed above. Regarding SIR, there is some concern about the reliability of components because access for maintenance is restricted. Extensive design and development are needed, and a full-scale first plant will probably be required before design certification is approved. In addition, numerous technical issues must be resolved to establish the systems' performance during normal operation and the adequacy of the safety features. PIUS incorporates much new technology and has only been demonstrated in laboratory experiments. There is also concern about the stability of the interface between its reactor coolant and the highly borated water in the surrounding pool. At a minimum, a reasonably large experiment that combines neutronics with thermal hydraulics would be required to alleviate this concern, but the Committee believes a full-scale first plant will probably be necessary.

The MHTGR needs an extensive R&D program to achieve commercial readiness in the early part of the next century. The construction and operation of a first plant would likely be required before design certification. Although there is worldwide experience with gas-cooled reactors, most of these reactors are sufficiently different from the MHTGR that much of this experience is not relevant to the technical uncertainties relating to the advanced reactor type.[21] Experience with the U.S. Fort St. Vrain reactor and the German THTR underscores the need to complete development and build a first plant to identify potential problems in a full-scale plant. The advanced gas-cooled reactor is claimed to have a unique safety feature in its encapsulated fuel particles. However, additional R&D would be needed to confirm that fission product containment in mass-produced, core quantity batches is achieved at severe accident temperatures of 1,800° to 2,000°C for extended periods with extremely high reliability. Oak Ridge National Laboratory estimates that data to confirm fuel performance will not be available before 1994.[Homan, 1989] The Committee also recognizes that the particle fuel concept of the MHTGR may lead to significant focus of regulatory safety inspection on the common fuel manufacturing facility, because of the required stringent quality of the approximately nine billion particles comprising each core and the maintenance of this regime throughout the operating lifetime of the manufacturing facility. Means to achieve assurance will have to be developed. The Committee believes that reliance on the defense-in-depth concept must be retained, and accurate evaluation of an advanced reactor's safety profile will require evaluation of a detailed design. Studies of accident scenarios should be continued, including the effect of air ingress accidents on the structural support of the core to assure that the core configuration does not change. Finally, the MHTGR does offer the unique capability of producing high-temperature process heat, but to achieve this potential, an extensive development program involving major components must be successfully completed.

The LMR already has an operating test bed reactor (EBR-II), an operating irradiation test facility (FFTF), and a well-framed program to develop LMR technology and demonstrate the integral fast reactor concept. Results to date are promising, and a modular plant design is being developed. This program is backed by a long history of LMR R&D in the United States. However, much R&D is still required. A federally funded program, including one or more first plants, will be required before any LMR concept would be

---

[21] If the MHTGR is selected for the new production reactor, substantial development funding for the military production version would also benefit the civilian version. The vendor association estimates that such benefits could amount to about $0.4 to $0.7 billion (i.e., a reduction from about $1 billion to $0.3 - $0.6 billion).[DOE, 1990]

accepted by U.S. utilities. For example, there is lingering concern about the use of sodium because of the possibility of sodium-water reactions and potential fire hazards, although relevant experience with such reactors to date has been positive on these points. There also is some safety concern about the large positive sodium void coefficient in some core designs, although the overall temperature and power coefficient are negative. The opacity of sodium makes the assurance of satisfactory in-vessel inspections and operations more difficult. An accident that produces significant core-wide boiling is very unlikely. In addition, containment is designed to withstand such an accident, including fuel melting.[Nuclear Power Assembly and ANS, 1990] Finally, *in situ* fuel reprocessing[22] must be demonstrated, and concerns about proliferation must be allayed. The economics of this technology, including costs of reprocessing facilities, can be demonstrated only after a first plant is built and operating.

Summary

The large evolutionary LWRs are judged to have the least development risk. The CANDU-3 reactor is farther along in design than the mid-sized LWRs with passive safety features. However, it has not entered NRC's design certification process. For these designs it is probable that a first plant will not be required for certification. However, the Committee believes that, while a prototype in the traditional sense will not be required, federal funding will likely be required for the first mid-sized LWR plant with passive safety features to be ordered. The remaining reactor technologies have significant development risk, and all will require a federally supported first plant.

Licensing

Discussion

The large evolutionary LWRs are furthest along in the design certification process. They clearly should be most amenable to efficient and predictable licensing and will very likely be the first to be certified. For the mid-sized LWRs with passive safety features, EPRI is working closely with the industry to help move the licensing process forward. These reactors are likely to be the next type certified.

---

[22] It is possible that centralized reprocessing may be selected instead of *in situ* reprocessing.

The CANDU reactor can probably be licensed in this century, although it is probably farther behind in the process than the mid-sized LWRs with passive safety features. Moreover, obtaining certification for the CANDU could require substantial additional work on the part of the developer because of great differences in Canadian and U.S. regulatory systems.[Ahearne, 1989]

The SIR and PIUS reactors are still farther behind in the licensing process, and much R&D would have to be done before they could apply for certification. However, these reactors appear to be certifiable eventually, although a first plant will probably be needed. With adequate funding to complete the development program, a demonstration plant for the MHTGR could be licensed slightly after the turn of the century, with certification following demonstration of successful operation. The LMR based on the integral fast reactor concept is still in a very early stage, with much new technology to be evaluated. Reprocessing and recycling will raise significant licensing issues. From the viewpoint of commercial licensing, it is far behind the evolutionary and mid-sized LWRs with passive safety features in having a commercial design available for review.

Summary

It would appear that the large evolutionary LWRs could obtain a NRC design certification as soon as the early to mid-1990s, and the mid-sized LWRs with passive safety features perhaps a little later, followed by CANDU. First plants will probably be required for the other reactor concepts, whose design certification would not be forthcoming until perhaps a decade or more later. The alternative R&D programs presented in Chapter 4 reflect these judgments.

Overall Assessment

The Committee's overall assessment of these technologies is that the large evolutionary LWRs and the mid-sized LWRs with passive safety features rank highest relative to the evaluation criteria. The evolutionary reactors could be ready for deployment by 2000, and the mid-sized could be ready for initial plant construction soon after 2000. The mature evolutionary designs would be available if significant new nuclear generating capacity should be needed before the mid-sized LWRs are ready. Both types of LWRs take advantage of the extensive experience with current reactors, yet they promise improvements in the most troublesome aspects of that experience (e.g., cost, schedule, and licensing). Determinants of the choice among these systems would be perceived financial risk and associated financial arrangements, capacity requirements, and availability of certified, standardized designs.

The heavy water reactor is also a mature design, and Canadian entry into the U.S. marketplace would give added insurance of adequate nuclear capacity if it is needed in the future. But the CANDU does not offer advantages sufficient to justify U.S. government assistance to initiate and conduct its licensing review.

The other LWR concepts (SIR and PIUS), the MHTGR, and the advanced LMR are believed to be considerably less mature and hence not likely to be deployed for commercial use in the United States until perhaps 2010 to 2025 or later, assuming their development proceeds. SIR and PIUS primarily offer safety benefits. The advanced gas-cooled reactor offers safety benefits and the potential of producing process heat. The advanced LMRs are also judged to offer benefits in their safety and in their ability to breed fuel should uranium resources become scarce. Their potential to alleviate some of the waste disposal problem for LWR fuel through actinide recycling is in such a preliminary stage that this feature is not considered justification for advancing the advanced LMR development program nor delaying waste repository schedules. The Committee judges that the MHTGR process heat capability is of little strategic significance compared with the LMR's potential for breeding. Based on information available at the time of the Committee's review, the Committee did not judge the safety benefits among the reactors discussed in this paragraph to be significantly different, and thus safety is not a discriminant. The development required for commercialization of any of these concepts is substantial.

The Committee's evaluations and overall assessment are summarized in Figure 3-12.

The Committee's major conclusions regarding the advanced reactor technologies flow from the above assessment. These conclusions are as follows:

1. Safety and cost are the most important characteristics for future nuclear power plants.
2. LWRs of the large evolutionary and the mid-sized advanced designs offer the best potential for competitive costs (in that order).
3. Safety benefits among all reactor types appear to be about equal at this stage in the design process. Safety must be achieved by attention to all failure modes and levels of design by a multiplicity of safety barriers and features. Consequently, in the absence of detailed engineering design and because of the lack of construction and operating experience with the actual concepts, vendor claims of safety superiority among conceptual designs cannot be substantiated.

Reactor Designation	Available Design Information	Evaluation Criteria: Safety	Economy	Market Suitability	Fuel Cycle	Safeguards & Physical Sec.	Maturity of Development	Licensing	Overall Assessment[1]
ABWR[a]	○	○	○	○	◐	◐	○	○	○
APWR[a]	○	○	○	○	◐	◐	○	○	○
SYS 80+[a]	○	○	○	○	◐	◐	○	○	○
AP 600[b]	◐	○	◐	○	◐	◐	◐	○	○
SBWR[b]	◐	○	◐	○	◐	◐	◐	○	○
CANDU	○	○	●	◐	◐	◐	○	◐	◐
SIR	◐	○	◐ [2]	●	◐	◐	●	●	●
MHTGR	◐	○	●	●	◐	◐	●	●	●
PIUS	●	○	◐ [2]	●	◐	◐	●	●	●
PRISM[c] LMR	●	○	● [3]	● [3]	○	◐	●	●	◐

Legend Rating: ○ high ◐ moderate ● low

Notes:
1 Overall assessment was mostly driven by market suitability.
2 Lack of design maturity results in great uncertainty relative to vendor cost projections.
3 Long-term economy and market potential could be high, depending on uranium resource availability.
a ABWR - *advanced boiling water reactor;* APWR - *advanced pressurized water reactor; and* SYS 80+ - *system 80+. Large evolutionary LWRs.*
b AP 600 - *advanced pressurized 600;* and SBWR - *simplified boiling water reactor. Mid-sized LWRs with passive safety features.*
c PRISM - *power reactor, innovative small module.*

FIGURE 3-12 Assessment of advanced reactor technologies.

This table is an attempt to summarize the Committee's qualitative rankings of selected reactor types ***against each other***, without reference either to an absolute standard or to the performance of any other energy resource options. This evaluation was based on the Committee's professional judgment.

4. LWRs can be deployed to meet electricity production needs for the first quarter of the next century:[23]

a. The evolutionary LWRs are further developed and, because of international projects, are most complete in design. They are likely to be the first plants certified by NRC. They are expected to be the first of the advanced reactors available for commercial use and could operate in the 2000 to 2005 time frame. Compared to current reactors, significant improvements in safety appear likely. Compared to recently completed high-cost reactors, significant improvements also appear possible in cost if institutional barriers are resolved. While little or no federal funding is deemed necessary to complete the process, such funding could accelerate the process.

b. Because of the large size and capital investment of evolutionary reactors, utilities that might order nuclear plants may be reluctant to do so. If nuclear power plants are to be available to a broader range of potential U.S. generators, the development of the mid-sized plants with passive safety features is important. These reactors are progressing in their designs, through DOE and industry funding, toward certification in the 1995 to 2000 time frame. The Committee believes such funding will be necessary to complete the process. While a prototype in the traditional sense will not be required, federal funding will likely be required for the first mid-sized LWR with passive safety features to be ordered.

c. Government incentives, in the form of shared funding or financial guarantees, would likely accelerate the next order for a light water plant. The Committee has not addressed what type of government assistance should be provided nor whether the first advanced light water plant should be a large evolutionary LWR or a mid-sized passive LWR.

5. The CANDU-3 reactor is relatively advanced in design but represents technology that has not been licensed in the United States. The Committee did not find compelling reasons for federal funding to the vendor to support the licensing.

6. SIR and PIUS, while offering potentially attractive safety features, are unlikely to be ready for commercial use until after 2010. This alone may limit their market potential. Funding priority for research on these reactor systems is considered by the Committee to be low.

7. MHTGRs also offer potential safety features and possible process heat applications that could be attractive in the market place. However, based on the extensive experience base with light water technology in the United States, the lack of success with commercial use of gas technology, the likely

---

[23] While this may lock the U.S. into LWR technology for the next 20+ years, the reasons for which are summarized in the following paragraphs, it does not discourage research and development of competitive technologies which may be needed later, as described in Chapter 4.

higher costs of this technology compared with the alternatives, and the substantial development costs that are still required before certification,[24] the Committee concluded that the MHTGR had a low market potential. The Committee considered the possibility that the MHTGR might be selected as the new tritium production reactor for defense purposes and noted the vendor association's estimated reduction in development costs for a commercial version of the MHTGR. However, the Committee concluded, for the reasons summarized above, that the commercial MHTGR should be given low priority for federal funding.

8. The LMR technology also provides enhanced safety features, but its uniqueness lies in the potential for extending fuel resources through breeding. While the market potential is low in the near term (before the second quarter of the next century), it could be an important long-term technology, especially if it can be demonstrated to be economic. The Committee believes that the LMR should have the highest priority for long-term nuclear technology development.

9. The problems of proliferation and physical security posed by the various technologies are different and require continued attention. Special attention will need to be paid to the LMR.

The above conclusions formed the basis for the formulation of alternative U.S. R&D programs in Chapter 4.

---

[24] The Gas Cooled Reactor Associates estimates that, if the MHTGR is selected as the new tritium production reactor, development costs for a commercial MHTGR could be reduced from about $1 billion to $0.3 - $0.6 billion.[DOE, 1990]

REFERENCES

ABB. 1989. PIUS, Presentation Material for Committee on Future Nuclear Power Development, ABB Atom. July 26, 1989.

AECL. Undated. CANDU 300 Technical Outline. Revision Seven. Document PPS-74-01010-003. Received 1989.

AECL Technologies. 1989. Presentation by AECL Technologies to the National Academy of Sciences' Committee on Future Nuclear Power Development Reactor Technologies Workshop and Accompanying Documents. Irvine, CA. August 23, 1989.

Ahearne, J. F. 1989. A Comparison Between Regulation of Nuclear Power in Canada and the United States. Progress in Nuclear Energy. 22:3. 1988. Received May 16, 1989.

American Physical Society. 1978. Reviews of Modern Physics. Report to the APS by the Study Group on Nuclear Fuel Cycles and Waste Management. 50:1. Part II. January.

Beckjord, E. 1989. Nuclear Regulatory Commission's Director of Research. Letter to Mary Ann Novak, Acting Assistant Secretary for Nuclear Energy. DOE. February 28, 1989.

Berglund, R.C. 1989. ALMR Design and Program Summary. General Electric (Presentation to Committee on Future Nuclear Power Development. August).

Bradbury, R., J. Longo, R. Strong, and M. Hayns. 1989. The Design Goals and Significant Features of the Safe Integral Reactor. Presented at the ANS 1989 Annual Meeting. Atlanta, Georgia. June 4-8, 1989.

Bredolt, Ulf et al. 1988. PIUS-The Next Generation Water Reactor. American Nuclear Society Conference: Safety of Next Generation Power Reactors. Seattle, Washington. May.

CE. 1989a. System 80+ ™ Advanced Light Water Reactor Technology Assessment, A Report to the Committee on Future U.S. Nuclear Power Development.

CE. 1989b. Simplified Passive Advanced Light Water Reactor Plant Program. Safe Integral Reactor (SIR) Technology Assessment. A Report to the Committee on Future U.S. Nuclear Power Development.

Chang, Y.I., M.J. Lineberry, L. Burris, and L.C. Waters. 1987. Nucl. Eng. Int. 32:23(November).

Chang, Y.I. 1989. Integral Fast Reactor Technology. Presentation to National Academy of Sciences Committee on Future Nuclear Power Development. Argonne National Laboratory. August 21-25, 1989

Chilk, S. J., Secretary, U.S. Nuclear Regulatory Commission. 1991. Memorandum for James M. Taylor, Executive Director for Operations. Subject: SECY-90-377. Requirements for Design Certification Under 10 CFR Part 52. February 15, 1991.

Chung, K., and G.A. Hazelrigg. 1989. Nuclear Power Technology: A Mandate for Change. Nuclear Technology. 88(November).

Collier, T.G., and G.F. Hewitt. 1987. Introduction to Nuclear Power. Hemisphere Pub. Corp.

DOE. 1990. HTGR. MHTGR Cost Reduction Study Report. DOE-HTGR-88512. Issued by Gas Cooled Reactor Associates. October.

Duncan, J. D. and R. J. McCandless. 1988. Safety of Next Generation Power Reactors. A Paper Presented at the ANS Topical Meeting. Seattle, WA. May 1-5, 1988.

EPRI. 1990. Advanced Light Water Reactor Utility Requirements Document. ALWR Policy and Summary of Top-Tier Requirements. 1(March).

EPRI. 1989a. Utility Industry Evaluation of the Modular High Temperature Gas-Cooled Reactor. August.

EPRI. 1989b. Technical Assessment Guide: Electricity Supply--1989. EPRI P-6587-L. 1:(September). Rev. 6. Special Report.

GA. 1989. Modular High Temperature Gas-Cooled Reactor Technology and Applications. A Presentation to the National Academy of Sciences' Committee on Future Nuclear Power Development. August 23 and 24, 1989.

Gas-Cooled Reactor Associates. 1987. A Utility/User Summary Assessment of the Modular High Temperature Gas-Cooled Reactor Conceptual Design. Nov. GCRA Report 87-011. Revision 1.

Gas-Cooled Reactor Associates. 1989. Eleventh International HTGR Conference Overview. News Letter. Summer.

GE Nuclear Energy. 1989. GE Advanced LWR Technology, Presentation to National Academy of Sciences Committee on Future Nuclear Power Development, August 22, 1989.

Griffith, J. D. 1988. Advanced Reactor Program Presentation to the National Research Council. U. S. Department of Energy Nuclear Research and Development Program. December.

Griffith, J., Associate Deputy Assistant Secretary for Reactor Systems Development and Technology, Office of Nuclear Energy, DOE. 1990. Letter to Norm Haller, National Research Council. December 14, 1990. Containing drawings and extracts from Section G.4.1 of Appendix G. Amendment 13 to the PRISM (ALMR) Preliminary Safety Information Document. May.

Hill, Phil. 1989. Germany Shuts Down Two New Nukes. Environmental Action. November/December. p. 17.

Hirata, K., M. Negishi, C. Matsumoto et al. 1989. Advanced Pressurized Water Reactor Plant. Nuclear Europe. 11-12.

Homan, F. 1989. Readiness of MHTGR Technology for Commercial Development. Presentation to the National Research Council. National Academy of Sciences. Committee on Future Nuclear Power Development. Irvine, CA. August 24, 1989.

International Atomic Energy Agency. 1990. Nuclear Power Reactors in the World. Reference Data Series No. 2. April Edition.

Kintner, E. E. 1989. Letter to Archie L. Wood. November 9, 1989.

Krüger, K. (AVR-FRG) and Cleveland, J. (ORNL). 1989. Loss-of-Coolant Accident Experiment at the AVR Gas-Cooled Reactor. Thermal-Hydraulic Aspects of Passive Safety and New Generation Reactors-III. Winter Meeting of the American Nuclear Society. Transactions of the American Nuclear Society. 60:735-736. November.

Lewis, H.W. 1978. Chairman. Risk Assessment Review Group Report to the U.S. Nuclear Regulatory Commission. NUREG/CR-0400. September.

McCutchan, D., T. van de Venne, and J. Cobian. 1989. Improved Availability and Operation in the Advanced Pressurized Water Reactor (APWR). Presented at the International Conference on Availability Improvements in Nuclear Power Plants. Madrid, Spain. April.

National Research Council. 1990. Confronting Climate Change. Strategies for Energy Research and Development. August.

National Research Council. 1984. A Review of the Swedish KBS-3 Plan for Final Storage of Spent Nuclear Fuel. National Academy Press. Washington, D.C.

National Research Council. 1983. A Study of the Isolation System for Geologic Disposal of Radioactive Wastes. National Academy Press. Washington, D.C.

NRC. 1990a. Advisory Committee on Reactor Safeguards. Letter to Chairman Carr. Subject: Review of NUREG-1150, "Severe Accident Risks: An Assessment for Five U.S. Nuclear Power Plants". November 15, 1990.

NRC. 1990b. Memorandum for Chairman Carr and Commissioners Rogers, Curtiss, and Remick from James M. Taylor. Executive Director for Operations. Subject: Response to Items 1 and 2 of the SRM of 10/02/90, "Support for Reviews of CANDU 3 and PIUS Designs." November 23, 1990.

NRC. 1990c. Memorandum for Chairman Carr and Commissioners Rogers, Curtiss, and Remick from James M. Taylor. Executive Director for Operations. Subject: Response to Items 1 and 2 of the SRM of 10/02/90, "Support for Reviews of CANDU 3 and PIUS Designs." December 17, 1990. Enclosure: Memorandum to A.T. Gody from J. G. Giitter. Subject: Summary of Working Meeting with Atomic Energy of Canada Limited Technologies. November 28, 1990.

NRC. 1991a. Transcript of 30th Meeting of the Advisory Committee on Nuclear Waste. Bethesda, Maryland. April 24, 1991.

NRC. 1991b. Transcript of meeting entitled Briefing on Progress of Design Certification Review and Implementation. Accompanying viewgraphs. June 12, 1991.

Nuclear Power Assembly and ANS. 1990. PRISM, the Plant Design Concept for the U.S. Advanced Liquid Metal Reactor Program. Paper given at the Nuclear Power Assembly in Washington, D.C. in May and ANS Conference in Nashville, Tennessee in June.

Nylan, A. et al. 1988. In Proc. of the Intersociety Energy Conversion Engineering Conference. Denver, CO. August (American Society Mech. Engr. New York). pp. 483-488.

Pigford, T.H. 1990. Department of Nuclear Engineering. University of California, Berkeley. Letter to Archie L. Wood. National Research Council. November 19, 1990. Transmitting copy of paper entitled "Actinide Burning and Waste Disposal". An Invited Review for the MIT International Conference on the Next Generation of Nuclear Power Technology. UCB-NE-4176. October 5, 1990.

Taylor, J. J.and K.E. Stahlkopf. 1988. (no title). Nuclear Engineering Design. 109:19(September-October).

Taylor, J. J. 1989. Improved and Safer Nuclear Power. Science. 244:318-325. April 21, 1989.

Till, C.E. 1989. The Liquid Metal Reactor. Overview of the Integral Fast Reactor Rationale and Basis for Its Development. Presentation to National Academy of Sciences Committee on Future Nuclear Power Development. Argonne National Laboratory. August 21-25, 1989.

Westinghouse Electric Corporation. 1989. Assessing the Merits of AP600 Advanced Reactor Technology for U.S. Electric Power Needs. Report to the Committee on Future U.S. Nuclear Power Development. (Energy Systems Business Unit, AP600 Program Office). August.

Williams, P.M., T.L. King, and J.N. Wilson. 1989. Draft Preapplication Safety Evaluation Report for the Modular High-Temperature Gas-Cooled Reactor. NUREG-1338. Division of Regulatory Applications. U.S. Nuclear Regulatory Commission. Washington, D.C. 20555. March.

Wolfe, B. and D.R. Wilkens. 1988. Improvements in Boiling Water Reactor Designs and Safety. Presented at the American Nuclear Society Topical Meeting. Seattle, WA. May 1-5, 1988.

Young, W. H., Assistant Secretary for Nuclear Energy, DOE. 1989. Letter to Eric Beckjord. Director. Office of Nuclear Regulatory Research. U.S. Nuclear Regulatory Commission. November 28, 1989. Enclosure entitled Summary - Containment Study for Modular High Temperature Gas-Cooled Reactor (MHTGR).

Young, W. H. Assistant Secretary for Nuclear Energy, DOE. Undated. Letter to Professor Thomas H. Pigford. (Letter was undated, but received by the Committee on March 5, 1991).

4

Federal Research and Development Alternatives

The Committee was asked to develop a set of federal research and development (R&D) alternatives to guide a future civilian nuclear power development program. The alternatives presented here are based on the Committee's evaluation of the full range of practical technologies for future nuclear plants; they reflect no comparative evaluation of non-nuclear options for R&D funding. The formulation of this set of alternatives reflects an assessment of the relevance of existing civilian reactor development facilities of the Department of Energy (DOE) and the need for any new facilities and is based on the Committee's evaluation described in Chapter 3.[1]

Three R&D alternatives are presented in this chapter following a summary of the current DOE programs. Funding requirements for the various elements of each alternative were estimated by the Committee based on DOE information. None of these alternatives addresses DOE's programs for high-level radioactive waste disposal. While demonstration reactors for these alternatives are identified, funding for final design and construction is not included.

---

[1] This chapter addresses alternative Department of Energy funding levels to support future civilian nuclear power development. This is consistent with the Senate Appropriations Committee Report 100-381 that formed the basis for this study (see Preface). The Committee did not attempt to assess (a) the ability (or willingness) of private industry to underwrite part or all of these R&D costs, (b) the appropriate federal role in either prototype or final development, or (c) comparisons between federal funding for civilian nuclear power programs and other energy related programs that compete for federal R&D resources.

CURRENT PROGRAMS

Funding Projections for Near-Term and Long-Term Technologies

During the five Fiscal Years 1992 through 1996 DOE has proposed to spend about $1.6 billion on R&D for civilian nuclear power. Funding projections are about $0.2 billion for the near-term reactor technologies and about $1.4 billion for the long-term reactor technologies, including about $0.7 billion for support facilities (discussed below).[Rohm, 1991] For the near term, advanced mid-sized light water reactors (LWR) with passive safety features are being developed in cooperation with the nuclear industry. DOE also is providing some assistance to the industry's development of large evolutionary LWRs. For the long term, DOE is currently funding the development of modular high-temperature gas-cooled reactors (MHTGR) and liquid metal reactors (LMR).

The funding projections for facilities include about $0.2 billion for shutdown of the Hanford fast flux test facility (FFTF) over the period FY 1992 through 1996. DOE has proposed to shut down the FFTF complex, although Congress has appropriated funds for continued operation in FY 1991.

Facilities Currently Supported by Department of Energy

No funds are presently provided for DOE test facilities to support the development of commercial LWRs, nor have any facilities been identified and requested for DOE funding by the nuclear industry. However, many DOE test facilities currently exist in support of the LMR development program. [Hunter, 1989] The most important of these facilities are:

1. The FFTF (located in Washington)--a large LMR designed for irradiation tests of multiple full-sized metallic or oxide fuel elements in realistic conditions. It also has the capability for testing fuels and materials for a wide range of fission and fusion concepts, including safety related experiments.
2. The experimental breeder reactor-II (EBR-II, located in Idaho)--a LMR, which serves as an irradiation test bed for metallic fuel elements of small modular reactors and as a test bed for safety experiments. Although the fuel element lengths are shorter than those envisioned for advanced commercial LMRs, EBR-II is a major element of the integral fast reactor (IFR) program.
3. The hot fuel examination facility/south (HFEF/S, located in Idaho)--used for support of the IFR program on metallic fuel.

4. The fuel manufacturing facility (FMF, located in Idaho)--used for demonstration of manufacturing technology for LMR fuel elements.
5. The transient reactor test facility (TREAT, located in Idaho)--a facility for transient tests of fuel elements and clusters of elements.
6. The zero power physics reactor (ZPPR, located in Idaho)--a critical test facility for neutronic physics tests of new core concepts.
7. The Energy Technology Engineering Center (ETEC, located in California)--facilities for development and testing of components.
8. The hot fuel examination facility/north (HFEF/N, located in Idaho)--a hot cell facility for examination of irradiated fuel.
9. The Hanford hot fuel examination facility (HFEF, located in Washington)--a hot cell facility for examination of irradiated fuel.

DOE does not have any major facilities to support exclusively the development of the MHTGR or Canadian heavy water reactor concepts. However, there is some support equipment used at Oak Ridge National Laboratory for studies of the high temperature behavior of fuel particles for gas reactors.

FEDERAL RESEARCH AND DEVELOPMENT ALTERNATIVES

Three alternative DOE R&D programs are presented. The Committee has concluded that each alternative would help retain nuclear power as an option for meeting U.S. electric energy requirements, albeit with significantly different long-term implications. These alternatives have progressively higher levels of funding. No consideration has been made of how funding for these three alternatives should compare to funding for other energy related programs that compete for federal R&D resources.

A key feature of the alternatives is a clear delineation between research activities and development activities. A properly formulated civilian nuclear energy program should include a continuing, broad-based research component aimed at identifying promising new concepts and at confirming the feasibility of critical features of concepts already identified. In contrast, the development component should identify and pursue only one or two concepts in recognition of the large commitment of resources necessary for successful development, first-plant demonstrations, and commercialization of any reactor concept.

The assumptions upon which the alternatives are based are presented first followed by the research elements common to all three policy alternatives.

Assumptions

Irradiation Test Capability

Nuclear reactor development requires evaluation and integration of mechanical, electrical, electronic, and neutronic design concepts. Among the most difficult, and the most time-consuming, to evaluate are the issues concerning fuel behavior, particularly new fuel concepts and under transient conditions. Therefore, the successful conduct of any new U.S. reactor development program in which new fuel and core materials are employed requires a versatile, reliable, high-temperature irradiation test capability. This capability is essential to the development of fuels and other in-reactor materials such as moderator, structural, and control materials. In addition, it will provide the means for studying fission product behavior in both normal and accident environments for a fuel design concept.

Federal irradiation test facilities contributed significantly to the development of materials technology in the naval reactors program.[DOE and DOD, 1988; Hewlett, 1974; Westinghouse, 1958] The naval irradiation test programs primarily utilized the irradiation test facilities of the materials test reactor (MTR), engineering test reactor (ETR), advanced test reactor (ATR), and the Canadian Chalk River test reactor with in-pile loops.

An irradiation capability should provide test-to-failure for sample materials and proof testing of prototypic materials and configurations essential to the development of reactor fuels. These tests should be carried out in an environment that matches, in all essential characteristics, the irradiation conditions to which the prospective fuel and cladding will be exposed. To achieve this fidelity of test-to-design, several desirable factors must be considered: (1) simultaneous achievement of representative fuel burnup and clad fluence; (2) test of full length fuel elements; and (3) test of prototypic-sized arrays of fuel pins under design conditions. These experiments should be done in facilities specifically designed to provide high neutron fluxes and proper energy spectrums so that the tests simultaneously test the fuel and its coating or clad under essentially prototypic conditions.

Consequently, the Committee's R&D alternatives are based on the following assumptions regarding irradiation test facilities. Such facilities are needed for both research programs and development programs. Research programs need facilities capable of screening materials to select candidate design materials and configurations. Development programs can benefit greatly from facilities capable of testing prototype configurations of design materials to full design conditions. If adequate irradiation facilities are unavailable, the reactor development program would have to accept the significant technical risk inherent in extrapolating from a firm, tested technical basis to desired design conditions. In a slower-paced program, such extrapolations may be minimized by use of successive plant sizes, albeit at increased future cost. However, in even a modestly aggressive program, extrapolations of fuel irradiation behavior are uncertain, and availability of adequate irradiation facilities becomes very important.

An alternative to a versatile irradiation test facility in the United States is contracting for irradiation services at foreign test reactors to achieve timely test data under prototypic test conditions. However, for LMR needs, the United States would have to negotiate with foreign owners of LMRs regarding the conduct of specific tests, which may or may not become available.[2]

Nuclear Regulatory Commission Research

For the advanced LWRs with passive safety features currently supported by DOE, a Nuclear Regulatory Commission (NRC) confirmatory research program is necessary to acquire the data and analytic tools required to make certification decisions on these designs. We have assumed this research will be provided. However, the NRC research budget has declined substantially in the 1980s, and NRC research funds may not be sufficient to support timely certification, currently scheduled to be 1992 for the large evolutionary LWRs and 1994 to 1995 for the mid-sized LWRs with passive safety features. We note that NRC's Nuclear Safety Research Review Committee (NSRRC) concluded in its report dated December 21, 1990

> The FYP [Five Year Plan] does not address specific research for advanced reactors, and the NSRRC recommends that RES [NRC's Office of Nuclear Regulatory Research] and the NRC give prompt attention to this important issue.
>
> The distribution of funds across the major program areas of the FYP is appropriate given current needs and available funds. It would be difficult, however, to sustain a viable nuclear safety research program to support the NRC if the current budgets are decreased. In fact, budget increases will be needed to address the requirements for new technologies under regulatory oversight.[Morrison, 1990]

---

[2] Out of pile research experiments on component materials behavior will also be of considerable importance.

Common Research Elements

In addition to R&D being done by industry, three research elements are common to all alternatives: (1) reactor research at federal facilities; (2) university research funding; and (3) support for research to improve the operational performance and to extend the lifetimes of existing nuclear plants in the United States. The Committee believes these elements must be funded adequately to retain nuclear power as an option for meeting U.S. electric energy requirements.

Reactor Research

Research activity is vital to provide fundamental understanding of fuel cycle aspects of technologies already identified and to develop new reactor concepts. For example, research on fuel cycle aspects of the metal fuelled fast LMR, which could be accomplished by operating HFEF/S and FMF, including evolution of key prototypic reactor design and safety features, would be funded. To provide irradiation testing, EBR-II would be operated. This would provide limited capability for LMR fuel testing and for safety research at a U.S. facility.

The Committee concluded that federal support for development of a commercial version of the MHTGR should be a low priority (see section entitled "Excluded Programs" later in this chapter). However, the fundamental design strategy of the MHTGR is based upon the integrity of the fuel ($\leq 1600^{o}C$) under operation and accident conditions. There are other potentially significant uses for such fuel, in particular, space propulsion.[Pierce, 1985; Powell and Horn, 1987; Powell and Botts, 1983; Powell et al., 1985] Consequently, the Committee believes that DOE should consider maintaining a coated fuel particle research program within that part of DOE focused on space reactors.

Additionally, research would explore reactor concepts not addressed in this report, materials in particular fuels, and design features that would otherwise not be examined in the United States. Future reactor development directions are vitally dependent on the ability of DOE to sustain such a component that can originate innovative ideas.

University Research

The second element, funding for research at universities, recognizes and exploits the potential of university academic programs to enhance DOE research by generating new technology and reactor concepts and to sustain the commercial nuclear power industry by producing technically educated graduates. A recent National Research Council report stresses the need for additional and continued support for nuclear engineering students and research faculty. Whereas the need remains strong, over the last decade, nuclear engineering academic programs at both the masters and undergraduate levels have declined in terms of (1) the number of students enrolling, (2) the number of schools offering such curricula, and (3) the number of research reactors on university campuses.[3][NAS, 1990]

Operational Performance Improvement and Plant Life Extension

The final common element in each alternative is the recommendation that DOE support research programs to improve the operational performance and investigate the means to achieve plant life extension of existing nuclear plants in the United States. The successful operation of existing plants is required to restore public confidence in the nuclear option, and the achievement of life extension will maintain the contribution of electricity production from existing nuclear plants substantially beyond their original licensed period.

Utilities find it difficult to justify R&D money as part of their rate base, and vendors have little incentive to carry out such research. These factors, a variety of other reasons, and an orientation to near-term "bottom-line" results limit the investments to levels lower than the Committee believes necessary. While NRC is doing some research in areas strictly related to its licensing responsibilities, the major share of the required R&D effort must come from DOE.

---

[3] Undergraduate senior enrollments in nuclear engineering programs decreased from 1,150 in 1978 to about 650 by 1988. Enrollments in masters programs also peaked in the late 1970s, at about 1,050 students, and steadily declined to about 650 students in 1988. Since 1982, however, student enrollment in doctoral programs has remained relatively steady at about 600.

The number of U.S. undergraduate nuclear engineering programs declined from 80 in 1975 to 57 in 1989.

Two decades ago, 76 U.S. university research reactors were operating. By 1987, 27 university research reactors were in operation at universities offering nuclear engineering degrees or options in nuclear engineering.[NAS, 1990]

Alternative Program 1: Light Water Reactor Development and Common Research Elements

Two very differing perspectives have been publicly espoused regarding the technology upon which to base retention of nuclear power as an option for meeting future U.S. electric energy requirements. The first is that the experience with light water technology has provided a foundation for this retention and should be pursued, incorporating improvements as they are developed. The second is that this experience has been sufficiently flawed by accidents and economic difficulties to warrant a shift to a completely new nuclear technology. The Committee believes the first perspective is correct because there is far greater experience with LWR plant and fuel design, construction, regulation, and operation; there is no need for substantial additional R&D; LWR technology can be deployed commercially with much shorter and more predictable schedules and costs; and it utilizes existing resources and infrastructures more effectively than other designs; all of which makes for less uncertainties with this technology for the near term than any alternative.

Chapter 3 presented the Committee's conclusion that LWRs are the most likely candidates for commercial purchase in the next 15 to 30 years. The Committee concluded that, of the advanced designs, the large, evolutionary LWRs will be the first to be certified in the United States. Work on these designs is mature, having been funded cooperatively by Japanese and U.S. industry. Consequently, the Committee sees no need for federal R&D funding for these concepts, although federal funding could accelerate the certification process. The Committee does see the evolutionary LWRs as the most likely to be available for purchase in the next few years, and has concluded that these reactors, if they meet their design goals, would satisfy safety and lifetime cost requirements. Therefore, if additional federal funding is required to meet unique NRC certification requirements, such funding would be consistent with retaining nuclear power as an option for the United States. Whether it should be included in a government program would require an analysis of industry's ability to fund such work and evaluation of the appropriate role of the federal government in assisting what are basically commercial products.

In addition to the evolutionary LWRs, the mid-sized LWR with passive safety features could serve to retain nuclear power as a U.S. option. Therefore, R&D Alternative 1 concentrates development funding on determining how improvements in the concept of a mid-sized LWR with passive safety features can be realized. The required single and integral scaled tests can be carried out in private industrial facilities. Hence, no new DOE test facilities are needed for mid-sized LWRs with passive safety features. However, these tests are vital for the development of such advanced reactors.

Consequently, the Committee judges that these tests will be required for each passive safety related system at a scale large enough to validate design criteria.

Long-term retention of the nuclear option would require assurance of long-term availability of economical fuel supply. The LMR, which can be designed to perform as a near breeder or a true breeder, is a concept whose introduction would provide this assurance. However, the date by which breeders will be needed is uncertain and depends principally on the rate of growth of nuclear power production in the United States and in other countries and on domestic and world natural uranium resources. Therefore, R&D Alternative 1 is based on the premise that the LMR development program, as opposed to research programs, for long-term needs could be deferred and initiated at a later date when the time frame of its need becomes more defined. Concurrent research on the LMR, which currently emphasizes the IFR concept, as well as investigation of other reactor basic research, is accomplished in Alternative 1 through the Reactor Research common element. (The IFR concept utilizes metal fuel processed in situ by pyrometallurgy.) In addition, the other two common research elements, university research and operational performance improvement and plant life extension of existing LWRs, are included in R&D Alternative 1.

The Committee also concludes that no first plant mid-sized LWR with passive safety features is likely to be certified and built without government incentives, in the form of shared funding or financial guarantees.[4] The Committee has not addressed what type of government financial assistance (if any) would be required nor what should be the specific type for the first LWR plant to be built. As a result, budget projections listed for the three alternatives include no allowances for federal investment in the actual licensing and construction of a reactor and therefore may be significantly less than what actually would be required. Whatever approach is used, the role of the industry and government must be explicit from the beginning.

In summary, this first alternative, which has the lowest cost, contains the three common elements, assumes that certification of evolutionary LWRs will not require further DOE funding, limits development to mid-sized LWRs with passive safety features, and maintains the LMR program as a research activity. The major facility for LMR irradiation research, EBR-II, would be retained under this option.

---

[4] DOE has estimated that lead plant engineering alone would require about $100 million of federal funding in the 1990s.[Griffith, 1990]

Introduction to Alternatives 2 and 3

As discussed in Chapter 2, the Committee believes that several factors other than reactor designs will strongly influence whether nuclear power will be retained as an option in the United States. We have concluded that, if nuclear power plants are to be ordered within the next few years, the evolutionary LWR will be available. If a later time, toward the end of this decade or early in the next, is the introduction point, R&D Alternative 1 is aimed at ensuring that the mid-sized LWRs with passive safety features also would be available.

Were the nuclear option to be chosen, and large scale, long-term deployment followed, uranium supplies at competitive prices would eventually become exhausted. This eventuality has been the fundamental and traditional basis for reactor development program strategies that have called for the future introduction of the breeder concept and, in past periods of optimism, for near-breeders to extend the time period before breeders were required.

The liquid metal cooled fission reactor is the most developed of potentially exploitable technologies and can be designed to operate over a range of conversion ratios. This flexibility, the positive LMR operating history, and the judgment arrived at in Chapter 3 that no other advanced concept has a discernable relative safety advantage led the Committee to conclude that the LMR should be the successor reactor to LWR technology. The estimate of the time for LMR commercial deployment should take into consideration the projected use of uranium and its consequent increase in price, the annual growth rate for U.S. electric generating capacity, and how that growth may be met. One recent National Research Council study[National Research Council, 1987] concluded that

> Depending on the extent of future use of light-water reactors, the total use and commitment of known U.S. uranium oxide resources (U_3O_8) at a price less than $200 per pound could occur as early as the year 2020; that circumstance would be more likely to occur between 2020 and 2045. Availability of global uranium supplies would delay this occurrence by about thirty years.

The effect of reduction in enrichment costs by successful introduction of advanced technologies such as AVLIS (atomic vapor laser isotope separation) were not considered in this study, but would tend to further extend these dates.

Based on the conclusion in the National Research Council study, introduction of the LMR breeder could occur as early as 2020 or as late as 2075. Other considerations are (1) the retention of an existing trained cadre of LMR engineers and scientists, together with existing facilities for LMR development, and (2) the time needed for development, prototype construction, and accumulation of sufficient prototype operation experience upon which to base a decision to commercialize the technology. The first factor is relevant to the latest target date while the second affects the earliest date.

The Committee believes that the development, construction, and operation phases require approximately fifteen-, ten-, and ten-year periods, respectively. This thirty-five year total period leads to an earliest date of 2025, but we have no information on which to estimate the impact of the retention factor. Hence, while the year 2025 is early in the range of the uranium cost scenarios, the Committee adopted the target date of 2025 as the earliest date of introduction.

Consequently, Alternatives 2 and 3 have been developed to provide R&D alternatives that would explicitly include development of the liquid metal breeder option rather than postponing the decision to a later date, as is the case under Alternative 1.

Alternative Program 2: Alternative 1 Plus Liquid Metal Reactor Development

The rationale for Alternative 2 is to maintain a national program that assures retention of the nuclear option for the longer term; this requires the continued availability of an economic nuclear fuel resource. The LMR employing a breeding cycle is the only assured means of providing this resource currently foreseen. Developments in competing nuclear and non-nuclear energy supply options, as well as technological progress on the LMR itself, will determine when it should be deployed. This R&D Alternative retains much of the existing program infrastructure and applies it to developing LMRs for possible commercial deployment by the year 2025. This target date allows the development program to proceed in a slower and less costly manner than is included in R&D Alternative 3.

In addition to funding development of the mid-sized LWRs and the common research elements embodied in Alternative 1, this alternative adds funding for development of LMRs for commercial deployment by the year 2025. This development program would encompass all conceptual and engineering design, component development, and testing that would allow the first LMR plant to be built and placed in service by the year 2015. Successful experience with this early (first) plant is needed to develop a sufficient economic, safety, and operational basis for commercial confidence in the design prior to beginning commercial deployment in 2025. The current program for development of an LMR would be expanded to begin more detailed design of a demonstration plant, but no funds are included for construction. Funding for constructing this first plant will need to be shared by government and industry.

This alternative would also include limited research to examine the feasibility of recycling actinides from LWR spent fuel, utilizing the LMR.

As in Alternative 1, Alternative 2 assumes completion of the planned shutdown of the FFTF, but provides irradiation capability through operation of EBR-II. The availability of the FFTF would reduce the magnitude of the extrapolations required in the conduct of the fuels and materials development programs for the LMR concept. However, in view of the proposed extended development schedule of this alternative and the recommendation to construct at least one first plant, the required fuel performance extrapolations might be tolerable. Consequently, the Committee believes that it is possible to explore structuring this program without the assumed availability of the FFTF, which would be shut down if the current DOE intent is fulfilled.

Alternative 2 does require that DOE test facilities in Idaho (TREAT and ZPPR) and ETEC continue to operate. They currently exist in support of the LMR program, but would be shut down under Alternative 1. With respect to ETEC, DOE should give priority to testing U.S. concepts for industrial components of an LMR. Additionally, cooperative development activities of mutual interest to Japanese or European designs should be pursued.

Some DOE facilities for examining hot fuel that are currently operating may not be needed for this alternative. The Committee did not attempt to determine which facilities should be retained to provide the necessary support for this level of research. The Committee believes that DOE should determine whether the Idaho HFEF/N or the Hanford HFEF should be closed down (or placed in standby).

Alternative Program 3: Alternative 1 Plus Accelerated Liquid Metal Reactor Development, Including Light Water Reactor Actinide Recycling Studies

This alternative continues the mid-sized LWR development program in the previous alternatives as well as the common elements. However, it expands the LMR program to make available the option of commercially deploying LMRs in 2015. The advancement in date from the year 2025 of Alternative 2 is adopted to reflect the earliest date the Committee believes is practically possible to ready the LMR for commercialization. This could be achieved by reductions of five years in both the required development phase and the prototype construction phase. The earlier date could become desirable if the LMR safety characteristics, the capacity of recycling of its own actinides, and a greater demand for breeding were to become publicly recognized as very desirable features. Alternative 3 includes accelerated development of the IFR concept. The FFTF is retained in this option for fuel irradiation testing.

This alternative also provides for investigation of the desirability and feasibility of recycling actinides from LWR spent fuel. A target date for determining the technical, economic, and political feasibility of actinide recycle for LWR spent fuel is 2005. Emphasis on actinide recycling is warranted if it can be shown to reduce significantly the time that waste in a geological repository needs to be isolated and that no adverse institutional or technical constraints to waste fuel management are introduced. However, in the Committee's view, it should be emphasized that actinide recycling studies are not a substitute for proceeding expeditiously to construct a high-level radioactive waste isolation facility. We also note that plutonium, the major alpha-active by-product of existing reactor operations, can be utilized in LMRs or, in mixed-oxide fuel, in LWRs. Of course, either requires reprocessing of spent fuel.

This R&D alternative will require an irradiation reactor facility for testing both fuel assemblies and composite fuel pins at as near to prototypic steady and transient conditions as possible and at accelerated testing times. These factors are most important for the LMR development proposed in this alternative because the development time cycle is to be accelerated and high fuel element burnup per fuel cycle is desired for recycling studies. Consequently, the Committee believes, unlike DOE [Griffith, Undated], that this development path would be far better pursued by maintaining the FFTF to support the LMR program. The FFTF is superior to EBR-II regarding the following valuable technical irradiation goals: (1) simultaneous match of irradiation damage to the clad based on neutron fluence with peak fuel burnup based on the energy averaged neutron flux for fission; (2) irradiation of prototypic fuel lengths with design peak-to-average axial power ratios to confirm axial fuel pin behavior particularly with respect to axial fuel motion under transient conditions; and (3) testing of prototypic sized fuel assemblies under prototypic irradiation, coolant temperature, and coolant velocity conditions. In fact, the Committee was informed by DOE that nine LMR test assemblies are currently being irradiated in the FFTF. Without the availability of the FFTF, significant extrapolation from test conditions to LMR design conditions will be required, although, as described in Alternative 2, this is possible. The amount of extrapolation will increase as the fuel is enriched in order to reduce testing times. Hence, the Committee retains the FFTF in this alternative.

EBR-II would remain operational until its current fuel irradiation and closed fuel cycle programs are completed. This transition is estimated to take 5 to 7 years.

Excluded Programs

None of the three alternatives presented contain funding for the development of the MHTGR. The Committee carefully reviewed the current state of the high-temperature gas-cooled reactor, including safety and economic considerations, as a technology for the generation of electricity and high-temperature gas. This assessment included the further R&D required, including attendant uncertainties, and the projected economics of the technology. The Committee concluded that no foreseeable commercial market exists for MHTGR-produced process heat, which is the unique strategic capability of the MHTGR. Further, as discussed in Chapter 3, the MHTGR does not offer demonstrable cost or safety advantages over the other concepts. Therefore, given the limited funds available for commercial nuclear power development and the desirability to focus and coalesce efforts behind light water and liquid metal technologies, no funds should be allocated for development of high-temperature gas-cooled reactor technology within the commercial nuclear power development budget of DOE.

The Committee also has concluded that no funds should be allocated for R&D on SIR, PIUS, or CANDU-3 (the other advanced reactors discussed in Chapter 3). However, the Committee has taken no position on private funding or international consortia for any of these reactor types, or for the MHTGR.

Costs of the Alternatives

The approximate annual costs for DOE of each of the above alternatives are depicted in Table 4-1. DOE programs for FY 1990 and FY 1991 are shown for comparison. (Note: The numbers in Table 4-1 are approximate; actual numbers would need to be developed by DOE or the Office of Management and Budget.)

The annual costs in Table 4-1 are average costs for the near term, about the next five years, for each alternative. These costs specifically include the costs to operate those facilities that have been identified as needed for each alternative. All of these alternatives will take considerably longer than five years to achieve commercialization of one or more reactor technologies. The life cycle costs of these alternatives, which must include any government contribution to first plant construction, remain to be developed.

TABLE 4-1 Near Term R&D Funding Required[a]

Programs:	Large Evolutionary Reactors	Mid-sized LWR Development	MHTGR Develop.	LMR		COMMON ELEMENTS				Total Per Year
				LMR		Reactor Research				
				Develop.	Facilities	Facilities	LMR/New Concepts	University Research	Performance and Life-extension of Existing Plants	
Base (Common Elements)	-	-	-	-	-	45[b]	20/5	5	10	85
R&D Alternatives										
1. LWR Development and Common Elements	-	30	-	-	-	45	20/5	5	10	115
2. Alternative 1+ LMR Development	-	30	-	20	25[c]	45	20/5	5	10	160[e]
3. Alternative 1+ Accelerated LMR Development, Including Actinide Recycling Studies	-	30	-	40	115[d]	45	20/5	5	10	270[e]
DOE Programs										
FY 1990	3	15	22	0[h]	<--------169------>		36	0	3	248[f]
FY 1991	7	20	19	0[h]	<--------176------>		38	0	4	264[g]

[a] Government costs for any first plant are not included.
[b] EBR-II, HFEF/S, and FMF
[c] TREAT, ZPPR, ETEC, and either HFEF/N or Hanford HFEF
[d] FFTF added.
[e] LMR demonstration plant funding is not included.
[f] Excludes $2 million for advanced LWR severe accident studies and $1 million for safety exchanges with the Soviet Union.
[g] Excludes $3 million for advanced LWR severe accident studies, $8 million for safety exchanges with the Soviet Union, and $3 million for early site permits.
[h] Assumes all LMR expenditures are for research, not development.

SOURCE: Committee estimates based on Department of Energy (Office of Civilian Reactor Development) information dated March 20, 1991 and subsequent communications.
NOTE: The numbers in Table 4-1 are approximate; actual numbers would need to be developed by DOE and/or the Office of Management and Budget.

Additional Considerations

Finally, for whatever alternative R&D program is selected, DOE and the nuclear industry must ensure (1) that the reactors designed and developed equal or exceed the top tier design safety requirements advocated by the Electric Power Research Institute (Table 3-2), and (2) that projected total lifetime generation costs are such that the electricity produced is competitive with electricity produced (or saved) by alternative baseload technologies.

For whatever alternative is selected, the Committee's budget projections are intended to include R&D funding to address concerns about the potential for the risk of diversion of sensitive nuclear materials. Special attention will need to be paid to the LMR.

It is the Committee's judgment that Alternative 2 should be followed because it:

- provides adequate support for the most promising near-term reactor technologies;
- provides sufficient support for LMR development to maintain the technical capabilities of the LMR R&D community;
- would support deployment of LMRs to breed fuel by the second quarter of the next century should that be needed; and
- would maintain a research program in support of both existing and advanced reactors.

SUMMARY

The alternative R&D programs developed in this chapter contain three common research elements: (1) reactor research using federal facilities; retained for the LMR are EBR-II, HFEF/S, and FMF. The Committee believes that DOE should consider maintaining a coated fuel particle research program within that part of DOE focused on space reactors; (2) university research programs; and (3) improved performance and life extension programs for existing U.S. nuclear power plants.

Alternative 1 adds funding to assist development of the mid-sized LWRs with passive safety features. Alternative 2 adds a LMR development program and associated facilities (TREAT, ZPPR, ETEC, and either HFEF/N or the Hanford HFEF). This alternative would also include limited research to examine the feasibility of recycling actinides from LWR spent fuel, utilizing the LMR. Finally, Alternative 3 adds FFTF and increases LMR funding to accelerate reactor and IFR fuel cycle development and examination of actinide recycle of LWR spent fuel.

None of the three alternatives contain funding for development of MHTGR, SIR, PIUS, or CANDU-3.

Significant analysis and research is required to assess both the technical and economic feasibility of recycling actinides from LWR spent fuel.

It is the Committee's judgment that Alternative 2 should be followed.

REFERENCES

DOE and DOD. 1988. A Review of the United States Naval Nuclear Propulsion Program. June.

Griffith, J.D., Associate Deputy Assistant Secretary for Reactor Systems Development and Technology, Office of Nuclear Energy, Department of Energy. 1990. Letter to Archie L. Wood, Director, Energy Engineering Board. July 16, 1990.

Griffith, J.D., Associate Deputy Assistant Secretary for Reactor Systems Development and Technology, Office of Nuclear Energy, Department of Energy. Undated. Letter to Archie L. Wood, Director, Energy Engineering Board. Received June 29, 1990. Forwarding Comments on the Draft Paper, The Need for the FFTF in the LMR Development Program.

Hewlett, R.G., and F. Duncan. 1974. Nuclear Navy, 1946-1962. University of Chicago Press.

Hunter, R.A. 1989. Presentation to the National Research Council on DOE-Nuclear Energy Test Facilities. Office of Facilities, Fuel Cycle and Test Programs, Department of Energy. October 19, 1989.

Morrison, D.L. 1990. Report on the U.S. Nuclear Regulatory Commission's Research Strategy from the Nuclear Safety Research Review Committee, December 21, 1990. (Forwarded via letter from David L. Morrison, Chairman, to Eric Beckjord, Director, Office of Nuclear Regulatory Research).

National Academy of Sciences. 1990. U.S. Nuclear Engineering Education: Status and Prospects, Committee on Nuclear Engineering Education. Washington, D.C.

National Research Council. 1987. Outlook for the Fusion Hybrid and Tritium-Breeding Fusion Reactors. National Academy Press. Washington, D.C.

Pierce, B. L. 1985. Gas Cooled Reactor Concepts for Space Applications. Space Nuclear Power Systems 1984. M. S. El-Genk and M. D. Hoover, eds. Orbit Book Co. Chapter 24. 191-196.

Powell, J. R., and T. E. Botts. 1983. Particle Bed Reactors and Related Concepts. Proc. Advanced Compact Reactor Systems. National Academy of Sciences. November 15-17, 1982. Washington, D.C.: National Academy Press. 95-153.

Powell, J. R., T. E. Botts, F. L. Horn, O. W. Lazareth, and J. L. Usher. 1985. SNUG: A Compact Particle Bed Reactor for the 400 to 4000 kWt Power Range. Space Nuclear Power Systems 1984. M. S. El-Genk and M. D. Hoover, eds. Orbit Book Co. Chapter 29. 239-248.

Powell, J. R., and F. L. Horn. 1987. High Power Density Reactors Based on Direct Cooled Particle Beds. Space Nuclear Power Systems 1984. M. S. El-Genk and M. D. Hoover, eds. Orbit Book Co. Chapter 39. 319-329.

Rohm, H.H. 1991. Letter to Norm Haller, National Research Council. Office of Civilian Reactor Development, Department of Energy. March 20, 1991.

Westinghouse Electric Corporation and Duquesne Light Company. 1958. The Shippingport Pressurized Water Reactor, Naval Reactors Branch, United States Atomic Energy Commission. Reading, Mass.:Addison Wesley Publishing Company.

5

Conclusions and Recommendations

The Committee was requested to analyze the technological and institutional alternatives to retain an option for future U.S. nuclear power deployment.

A premise of the Senate report directing this study is "that nuclear fission remains an important option for meeting our electric energy requirements and maintaining a balanced national energy policy." The Committee was not asked to examine this premise, and it did not do so. The Committee consisted of members with widely ranging views on the desirability of nuclear power. Nevertheless, all members approached the Committee's charge from the perspective of what would be necessary if we are to retain nuclear power as an option for meeting U.S. electric energy requirements, without attempting to achieve consensus on whether or not it should be retained. The Committee's conclusions and recommendations should be read in this context.

The Committee's review and analyses have been presented in previous chapters. Here the Committee consolidates the conclusions and recommendations found in the previous chapters and adds some additional conclusions and recommendations based upon some of the previous statements. The Committee also includes some conclusions and recommendations that are not explicitly based upon the earlier chapters but stem from the considerable experience of the Committee members.

Most of the following discussion contains conclusions. There also are a few recommendations. Where the recommendations appear they are identified as such by bold italicized type.

GENERAL CONCLUSIONS

In 1989, nuclear plants produced about 19 percent of the United States' electricity, 77 percent of France's electricity, 26 percent of Japan's electricity, and 33 percent of West Germany's electricity. However, expansion of commercial nuclear energy has virtually halted in the United States. In other countries, too, growth of nuclear generation has slowed or stopped. The reasons in the United States include reduced growth in demand for electricity, high costs, regulatory uncertainty, and public opinion. In the United States, concern for safety, the economics of nuclear power, and waste disposal issues adversely affect the general acceptance of nuclear power.

Electricity Demand

Estimated growth in summer peak demand for electricity in the United States has fallen from the 1974 projection of more than 7 percent per year to a relatively steady level of about 2 percent per year. Plant orders based on the projections resulted in cancellations, extended construction schedules, and excess capacity during much of the 1970s and 1980s. The excess capacity has diminished in the past five years, and ten year projections (at approximately 2 percent per year) suggest a need for new capacity in the 1990s and beyond. To meet near-term anticipated demand, bidding by non-utility generators and energy efficiency providers is establishing a trend for utilities acquiring a substantial portion of this new generating capacity from others. Reliance on non-utility generators does not now favor large scale baseload technologies.

Nuclear power plants emit neither precursors to acid rain nor gases that contribute to global warming, like carbon dioxide. Both of these environmental issues are currently of great concern. New regulations to address these issues will lead to increases in the costs of electricity produced by combustion of coal, one of nuclear power's main competitors. Increased costs for coal-generated electricity will also benefit alternate energy sources that do not emit these pollutants.

Costs

Major deterrents for new U.S. nuclear plant orders include high capital carrying charges, driven by high construction costs and extended construction times, as well as the risk of not recovering all construction costs.

Construction Costs

Construction costs are hard to establish, with no central source, and inconsistent data from several sources. Available data show a wide range of costs for U.S. nuclear plants, with the most expensive costing three times more (in dollars per kilowatt electric) than the least expensive in the same year of commercial operation. In the post-Three Mile Island era, the cost increases have been much larger. Considerable design modification and retrofitting to meet new regulations contributed to cost increases. From 1971 to 1980, the most expensive nuclear plant (in constant dollars) increased by 30 percent. The highest cost for a nuclear plant beginning commercial operation in the United States was twice as expensive (in constant dollars) from 1981 to 1984 as it was from 1977 to 1980.

Construction Time

Although plant size also increased, the average time to construct a U.S. nuclear plant went from about 5 years prior to 1975 to about 12 years from 1985 to 1989. U.S. construction times are much longer than those in other major nuclear countries, except for the United Kingdom. Over the period 1978 to 1989, the U.S. average construction time was nearly twice that of France and more than twice that of Japan.

Prudency

Billions of dollars in disallowances of recovery of costs from utility ratepayers have made utilities and the financial community leery of further investments in nuclear power plants. During the 1980s, rate base disallowances by state regulators totaled about $14 billion for nuclear plants, but only about $0.7 billion for non-nuclear plants.

Operation

Operation and maintenance (O&M) costs for U.S. nuclear plants have increased faster than for coal plants. Over the decade of the 1980s, U.S. nuclear O&M-plus-fuel costs grew from nearly half to about the same as those for fossil fueled plants, a significant shift in relative advantage.

Performance

On average, U.S. nuclear plants have poorer capacity factors compared to those of plants in other Organization for Economic Cooperation and Development (OECD) countries. On a lifetime basis, the United States is barely above 60 percent capacity factor, while France and Japan are at 68 percent, and West Germany is at 74 percent. Moreover, through 1988 12 U.S. plants were in the bottom 22. However, some U.S. plants do very well: 3 of the top 22 OECD plants through 1988 were U.S. U.S. plants averaged 65 percent in 1988, 63 percent in 1989, and 68 percent in 1990.

Except for capacity factors, the performance indicators of U.S. nuclear plants have improved significantly over the past several years. If the industry is to achieve parity with the operating performance in other countries, it must carefully examine its failure to achieve its own goal in this area and develop improved strategies, including better management practices. Such practices are important if the generators are to develop confidence that the new generation of plants can achieve the higher load factors estimated by the vendors.

Public Attitudes

There has been substantial opposition to new plants. The failure to solve the high-level radioactive waste disposal problem has harmed nuclear power's public image. It is the Committee's opinion, based upon our experience, that, more recently, an inability of states, that are members of regional compact commissions, to site low-level radioactive waste facilities has also harmed nuclear power's public image.

Several factors seem to influence the public to have a less than positive attitude toward new nuclear plants:

- no perceived urgency for new capacity;
- nuclear power is believed to be more costly than alternatives;
- concerns that nuclear power is not safe enough;
- little trust in government or industry advocates of nuclear power;
- concerns about the health effects of low-level radiation;
- concerns that there is no safe way to dispose of high-level waste; and
- concerns about proliferation of nuclear weapons.

The Committee concludes that the following would improve public opinion of nuclear power:

- a recognized need for a greater electrical supply that can best be met by large plants;
- economic sanctions or public policies imposed to reduce fossil fuel burning;
- maintaining the safe operation of existing nuclear plants and informing the public;
- providing the opportunity for meaningful public participation in nuclear power issues, including generation planning, siting, and oversight;
- better communication on the risk of low-level radiation;
- resolving the high-level waste disposal issue; and
- assurance that a revival of nuclear power would not increase proliferation of nuclear weapons.

Safety

As a result of operating experience, improved O&M training programs, safety research, better inspections, and productive use of probabilistic risk analysis, safety is continually improved. The Committee concludes that the risk to the health of the public from the operation of current reactors in the United States is very small. In this fundamental sense, current reactors are safe. However, a significant segment of the public has a different perception and also believes that the level of safety can and should be increased. The development of advanced reactors is in part an attempt to respond to this public attitude.

Institutional Changes

The Committee believes that large-scale deployment of new nuclear power plants will require significant changes by both industry and government.

Industry

One of the most important factors affecting the future of nuclear power in the United States is its cost in relation to alternatives and the recovery of these capital and operating charges through rates that are charged for the electricity produced. Chapter 2 of this report deals with these issues in some detail. As stated there, the industry must develop better methods for managing the design and construction of nuclear plants. Arrangements among the participants that would assure timely, economical, and high-quality construction of new nuclear plants, the Committee believes, will be prerequisites to an adequate degree of assurance of capital cost recovery from state regulatory authorities in advance of construction. The development of state prudency laws also can provide a positive response to this issue.

The Committee and others are well aware of the increases in nuclear plant construction and operating costs over the last 20 years and the extension of plant construction schedules over this same period.[1] The Committee believes there are many reasons for these increases but is unable to disaggregate the cost effect among these reasons with any meaningful precision.

Like others, the Committee believes that the financial community and the generators must both be satisfied that significant improvements can be achieved before new plants can be ordered. In addition, the Committee believes that greater confidence in the control of costs can be realized with plant designs that are more nearly complete before construction begins, plants that are easier to construct, use of better construction and management methods, and business arrangements among the participants that provide stronger incentives for cost-effective, timely completion of projects.

It is the Committee's opinion, based upon our experience, that the principal participants in the nuclear industry--utilities, architect-engineers, and suppliers--should begin now to work out the full range of contractual arrangements for advanced nuclear power plants. Such arrangements would

---

[1] See discussion of costs and construction schedules in Chapter 2.

increase the confidence of state regulatory bodies and others that the principal participants in advanced nuclear power plant projects will be financially accountable for the quality, timeliness, and economy of their products and services.

Inadequate management practices have been identified at some U.S. utilities, large and small, public and private. Because of the high visibility of nuclear power and the responsibility for public safety, a consistently higher level of demonstrated utility management practices is essential before the U.S. public's attitude about nuclear power is likely to improve.

Over the past decade, utilities have steadily strengthened their ability to be responsible for the safety of their plants. Their actions include the formation and support of industry institutions, including the Institute of Nuclear Power Operations (INPO). Self-assessment and peer oversight through INPO are acknowledged to be strong and effective means of improving the performance of U.S. nuclear power plants. The Committee believes that such industry self-improvement, accountability, and self-regulation efforts improve the ability to retain nuclear power as an option for meeting U.S. electric energy requirements. The Committee encourages industry efforts to reduce reliance on the adversarial approach to issue resolution.

It is the Committee's opinion, based upon our experience, that the nuclear industry should continue to take the initiative to bring the standards of every American nuclear plant up to those of the best plants in the United States and the world. Chronic poor performers should be identified publicly and should face the threat of insurance cancellations. Every U.S. nuclear utility should continue its full-fledged participation in INPO; any new operators should be required to become members through insurance prerequisites or other institutional mechanisms.

Standardization. The Committee views a high degree of standardization as very important for the retention of nuclear power as an option for meeting U.S. electric energy requirements. There is not a uniformly accepted definition of standardization. The industry, under the auspices of the Nuclear Power Oversight Committee, has developed a position paper on standardization that provides definitions of the various phases of standardization and expresses an industry commitment to standardization. The Committee believes that a strong and sustained commitment by the principal participants will be required to realize the potential benefits of standardization (of families of plants) in the diverse U.S. economy. It is the Committee's opinion, based upon our experience, that the following will be necessary:

• Families of standardized plants will be important for ensuring the highest levels of safety and for realizing the potential economic benefits of new nuclear plants. Families of standardized plants will allow standardized approaches to plant modification, maintenance, operation, and training.

• Customers, whether utilities or other entities, must insist on standardization before an order is placed, during construction, and throughout the life of the plant.

• Suppliers must take standardization into account early in planning and marketing. Any supplier of standardized units will need the experience and resources for a long-term commitment.

• Antitrust considerations will have to be properly taken into account to develop standardized plants.

Nuclear Regulatory Commission

An obstacle to continued nuclear power development has been the uncertainties in the Nuclear Regulatory Commission's (NRC) licensing process. Because the current regulatory framework was mainly intended for light water reactors (LWR) with active safety systems and because regulatory standards were developed piecemeal over many years, without review and consolidation, the regulations should be critically reviewed and modified (or replaced with a more coherent body of regulations) for advanced reactors of other types. ***The Committee recommends that NRC comprehensively review its regulations to prepare for advanced reactors, in particular, LWRs with passive safety features. The review should proceed from first principles to develop a coherent, consistent set of regulations.***

The Committee concludes that NRC should improve the quality of its regulation of existing and future nuclear power plants, including tighter management controls over all of its interactions with licensees and consistency of regional activities. Industry has proposed such to NRC.

The Committee encourages efforts by NRC to reduce reliance on the adversarial approach to issue resolution. ***The Committee recommends that NRC encourage industry self-improvement, accountability, and self-regulation initiatives.*** While federal regulation plays an important safety role, it must not be allowed to detract from or undermine the accountability of utilities and their line management organizations for the safety of their plants.

It is the Committee's expectation that economic incentive programs instituted by state regulatory bodies will continue for nuclear power plant operators. Properly formulated and administered, these programs should improve the economic performance of nuclear plants, and they may also enhance safety. However, they do have the potential to provide incentives counter to safety. The Committee believes that such programs should focus on economic incentives and avoid incentives that can directly affect plant safety. On July 18, 1991 NRC issued a Nuclear Regulatory Commission Policy Statement which expressed concern that such incentive programs may adversely affect safety and commits NRC to monitoring such programs. A joint industry/state study of economic incentive programs could help assure that such programs do not interfere with the safe operation of nuclear power plants.

It is the Committee's opinion, based upon our experience, that NRC should continue to exercise its federally mandated preemptive authority over the regulation of commercial nuclear power plant safety if the activities of state government agencies (or other public or private agencies) run counter to nuclear safety. Such activities would include those that individually or in the aggregate interfere with the ability of the organization with direct responsibility for nuclear plant safety (the organization licensed by the Commission to operate the plant) to meet this responsibility. ***The Committee urges close industry-state cooperation in the safety area.***

It is also the Committee's opinion, based upon our experience, that the industry must have confidence in the stability of NRC's licensing process. Suppliers and utilities need assurance that licensing has become and will remain a manageable process that appropriately limits the late introduction of new issues.

It is likely that, if the possibility of a second hearing before a nuclear plant can be authorized to operate is to be reduced or eliminated, legislation will be necessary. The nuclear industry is convinced that such legislation will be required to increase utility and investor confidence to retain nuclear power as an option for meeting U.S. electric energy requirements. The Committee concurs.

It is the Committee's opinion, based upon our experience, that potential nuclear power plant sponsors must not face large unanticipated cost increases as a result of mid-course regulatory changes, such as backfits. NRC's new licensing rule, 10 CFR Part 52, provides needed incentives for standardized designs.

Industry and the Nuclear Regulatory Commission

The U.S. system of nuclear regulation is inherently adversarial, but mitigation of unnecessary tension in the relations between NRC and its nuclear power licensees would, in the Committee's opinion, improve the regulatory environment and enhance public health and safety. Thus, the Committee commends the efforts by both NRC and the industry to work more cooperatively together and encourages both to continue and strengthen these efforts.

Department of Energy

Lack of resolution of the high-level waste problem jeopardizes future nuclear power development. The Committee believes that the legal status of the Yucca Mountain site for a geologic repository should be resolved soon, and that the Department of Energy's (DOE) program to investigate this site should be continued. In addition, a contingency plan must be developed to store high-level radioactive waste in surface storage facilities pending the availability of the geologic repository.

Environmental Protection Agency

The problems associated with establishing a high-level waste site at Yucca Mountain are exacerbated by the requirement that, before operation of a repository begins, DOE must demonstrate to NRC that the repository will perform to standards established by the Environmental Protection Agency (EPA). NRC's staff has strongly questioned the workability of these quantitative requirements, as have the National Research Council's Radioactive Waste Management Board and others. The Committee concludes that the EPA standard for disposal of high-level waste will have to be reevaluated to ensure that a standard that is both adequate and feasible is applied to the geologic waste repository.

Administration and Congress

The Price-Anderson Act will expire in 2002. The Committee sought to discover whether or not such protection would be required for advanced reactors. The clear impression the Committee received from industry representatives was that some such protection would continue to be needed, although some Committee members believe that this was an expression of desire rather than of need. At the very least, renewal of Price-Anderson in 2002 would be viewed by the industry as a supportive action by Congress and would eliminate the potential disruptive effect of developing alternative liability arrangements with the insurance industry. Failure to renew Price-Anderson in 2002 would raise a new impediment to nuclear power plant orders as well as possibly reduce an assured source of funds to accident victims.

Other

The Committee believes that the National Transportation Safety Board (NTSB) approach to safety investigations, as a substitute for the present NRC approach, has merit. In view of the infrequent nature of the activities of such a committee, it may be feasible for it to be established on an ad hoc basis and report directly to the NRC chairman. ***Therefore, the Committee recommends that such a small safety review entity be established.*** Before the establishment of such an activity, its charter should be carefully defined, along with a clear delineation of the classes of accidents it would investigate. Its location in the government and its reporting channels should also be specified. The function of this group would parallel those of NTSB. Specifically, the group would conduct independent public investigations of serious incidents and accidents at nuclear power plants and would publish reports evaluating the causes of these events. This group would have only a small administrative structure and would bring in independent experts, including those from both industry and government, to conduct its investigations.

It is the Committee's opinion, based upon our experience, that responsible arrangements must be negotiated between sponsors and economic regulators to provide reasonable assurances of complete cost recovery for nuclear power plant sponsors. Without such assurances, private investment capital is not likely to flow to this technology.

In Chapter 2, the Committee addressed the non-recovery of utility costs in rate proceedings and concluded that better methods of dealing with this issue must be established. The Committee was impressed with proposals for periodic reviews of construction progress and costs--"rolling prudency" determinations--as one method for managing the risks of cost recovery. The Committee believes that enactment of such legislation could remove much of the investor risk and uncertainty currently associated with state regulatory treatment of new power plant construction, and could therefore help retain nuclear power as an option for meeting U.S. electric energy requirements.

On balance, however, unless many states adopt this or similar legislation, it is the Committee's view that substantial assurances probably cannot be given, especially in advance of plant construction, that all costs incurred in building nuclear plants will be allowed into rate bases.

The Committee notes the current trend toward economic deregulation of electric power generation. It is presently unclear whether this trend is compatible with substantial additions of large-scale, utility-owned, baseload generating capacity, and with nuclear power plants in particular.

It is the Committee's opinion, based upon our experience, that regional low-level radioactive waste compact commissions must continue to establish disposal sites.

Summary

The institutional challenges are clearly substantial. If they are to be met, the Committee believes that the Federal government must decide, as a matter of national policy, whether a strong and growing nuclear power program is vital to the economic, environmental, and strategic interests of the American people. Only with such a clearly stated policy, enunciated by the President and backed by the Congress through appropriate statutory changes and appropriations, will it be possible to effect the institutional changes necessary to return the flow of capital and human resources required to properly employ this technology.

Alternative Reactor Technologies

Advanced reactors are now in design or development. They are being designed to be simpler, and, if design goals are realized, these plants will be safer than existing reactors. The design requirements for the advanced reactors are more stringent than the NRC safety goal policy. If final safety designs of advanced reactors, and especially those with passive safety features, are as indicated to this Committee, an attractive feature of them should be the significant reduction in system complexity and corresponding improvement in operability. While difficult to quantify, the benefit of improvements in the operator's ability to monitor the plant and respond to system degradations may well equal or exceed that of other proposed safety improvements.

The reactor concepts assessed by the Committee were the large evolutionary LWRs, the mid-sized LWRs with passive safety features,[2] the Canadian deuterium uranium (CANDU) heavy water reactor, the modular high-temperature gas-cooled reactor (MHTGR), the safe integral reactor (SIR), the process inherent ultimate safety (PIUS) reactor, and the liquid metal reactor (LMR). The Committee developed the following criteria for comparing these reactor concepts:

---

[2] The term "passive safety features" refers to the use of gravity, natural circulation, and stored energy to provide essential safety functions in such LWRs.

- safety in operation;
- economy of construction and operation;
- suitability for future deployment in the U.S. market;
- fuel cycle and environmental considerations;
- safeguards for resistance to diversion and sabotage;
- technology risk and development schedule; and
- amenability to efficient and predictable licensing.

With regard to advanced designs, the Committee reached the following conclusions.

Large Evolutionary Light Water Reactors

The large evolutionary LWRs offer the most mature technology. The first standardized design to be certified in the United States is likely to be an evolutionary LWR. The Committee sees no need for federal research and development (R&D) funding for these concepts, although federal funding could accelerate the certification process.

Mid-sized Light Water Reactors with Passive Safety Features

The mid-sized LWRs with passive safety features are designed to be simpler, with modular construction to reduce construction times and costs, and to improve operations. They are likely the next to be certified.

Because there is no experience in building such plants, cost projections for the first plant are clearly uncertain. To reduce the economic uncertainties it will be necessary to demonstrate the construction technology and improved operating performance. These reactors differ from current reactors in construction approach, plant configuration, and safety features. These differences do not appear so great as to require that a first plant be built for NRC certification. While a prototype in the traditional sense will not be required, the Committee concludes that no first-plant mid-sized LWR with passive safety features is likely to be certified and built without government incentives, in the form of shared funding or financial guarantees.

CANDU Heavy Water Reactor

The Committee judges that the CANDU ranks below the advanced mid-sized LWRs in market potential. The CANDU-3 reactor is farther along in design than the mid-sized LWRs with passive safety features. However, it has not entered NRC's design certification process. Commission requirements are complex and different from those in Canada so that U.S. certification could be a lengthy process. However, the CANDU reactor can probably be licensed in this century.

The heavy water reactor is a mature design, and Canadian entry into the U.S. marketplace would give added insurance of adequate nuclear capacity if it is needed in the future. But the CANDU does not offer advantages sufficient to justify U.S. government assistance to initiate and conduct its licensing review.

Modular High-Temperature Gas-Cooled Reactor

The MHTGR posed a difficult set of questions for the Committee. U.S. and foreign experience with commercial gas-cooled reactors has not been good. A consortium of industry and utility people continue to promote federal funding and to express interest in the concept, while none has committed to an order.

The reactor, as presently configured, is located below ground level and does not have a conventional containment. The basic rationale of the designers is that a containment is not needed because of the safety features inherent in the properties of the fuel.

However, the Committee was not convinced by the presentations that the core damage frequency for the MHTGR has been demonstrated to be low enough to make a containment structure unnecessary. The Oak Ridge National Laboratory estimates that data to confirm fuel performance will not be available before 1994. The Committee believes that reliance on the defense-in-depth concept must be retained, and accurate evaluation of safety will require evaluation of a detailed design.

A demonstration plant for the MHTGR could be licensed slightly after the turn of the century, with certification following demonstration of successful operation. The MHTGR needs an extensive R&D program to achieve commercial readiness in the early part of the next century. The construction and operation of a first plant would likely be required before design certification. Recognizing the opposite conclusion of the MHTGR proponents, the Committee was not convinced that a foreseeable commercial market exists for MHTGR-produced process heat, which is the unique strategic capability of the MHTGR. Based on the Committee's view on containment requirements, and the economics and technology issues, the Committee judged the market potential for the MHTGR to be low.

The Committee believes that no funds should be allocated for development of high-temperature gas-cooled reactor technology within the commercial nuclear power development budget of DOE.

Safe Integral Reactor and Process Inherent Ultimate Safety Reactor

The other advanced light water designs the Committee examined were the United Kingdom and U.S. SIR and the Swedish PIUS reactor.

The Committee believes there is no near-term U.S. market for SIR and PIUS. The development risks for SIR and PIUS are greater than for the other LWRs and CANDU-3. The lack of operational and regulatory experience for these two is expected to significantly delay their acceptance by utilities. SIR and PIUS need much R&D, and a first plant will probably be required before design certification is approved.

The Committee concluded that no Federal funds should be allocated for R&D on SIR or PIUS.

Liquid Metal Reactor

LMRs offer advantages because of their potential ability to provide a long-term energy supply through a nearly complete use of uranium resources. Were the nuclear option to be chosen, and large scale deployment follow, at some point uranium supplies at competitive prices might be exhausted. Breeder reactors offer the possibility of extending fissionable fuel supplies well past the next century. In addition, actinides, including those from LWR spent fuel, can undergo fission without significantly affecting performance of an advanced LMR, transmuting the actinides to fission products, most of which, except for technetium, carbon, and some others of little import, have half-lives very much shorter than the actinides. (Actinides are among the materials of greatest concern in nuclear waste disposal beyond about 300 years.) However, substantial further research is required to establish (1) the technical and the economic feasibility of recycling in LMRs actinides recovered from LWR spent fuel, and (2) whether high-recovery recycling of transuranics and their transmutation can, in fact, benefit waste disposal. Assuming success, it would still be necessary to dispose of high-level waste, although the waste would largely consist of significantly shorter-lived fission products. Special attention will be necessary to ensure that the LMR's reprocessing facilities are not vulnerable to sabotage or to theft of plutonium.

The unique property of the LMR, fuel breeding, might lead to a U.S. market, but only in the long term. From the viewpoint of commercial licensing, it is far behind the evolutionary and mid-sized LWRs with passive safety features in having a commercial design available for review. A federally funded program, including one or more first plants, will be required before any LMR concept would be accepted by U.S. utilities.

Net Assessment

The Committee could not make any meaningful quantitative comparison of the relative safety of the various advanced reactor designs. The Committee believes that each of the concepts considered can be designed and operated to meet or closely approach the safety objectives currently proposed for future, advanced LWRs. The different advanced reactor designs employ different mixes of active and passive safety features. The Committee believes that there currently is no single optimal approach to improved safety. Dependence on passive safety features does not, of itself, ensure greater safety. The Committee believes that a prudent design course retains the historical defense-in-depth approach.

The economic projections are highly uncertain, first, because past experience suggests higher costs, longer construction times, and lower availabilities than projected and, second, because of different assumptions and levels of maturity among the designs. The Electric Power Research Institute (EPRI) data, which the Committee believes to be more reliable than that of the vendors, indicate that the large evolutionary LWRs are likely to be the least costly to build and operate on a cost per kilowatt electric or kilowatt hour basis, while the high-temperature gas-cooled reactors and LMRs are likely to be the most expensive. EPRI puts the mid-sized LWRs with passive safety features between the two extremes.

Although there are definite differences in the fuel cycle characteristics of the advanced reactors, fuel cycle considerations did not offer much in the way of discrimination among reactors, nor did safeguards and security considerations, particularly for deployment in the United States. However, the CANDU (with on-line refueling and heavy water) and the LMR (with reprocessing) will require special attention to safeguards.

SIR, MHTGR, PIUS, and LMR are not likely to be deployed for commercial use in the United States, at least within the next 20 years. The development required for commercialization of any of these concepts is substantial.

It is the Committee's overall assessment that the large evolutionary LWRs and the mid-sized LWRs with passive safety features rank highest relative to the Committee's evaluation criteria. The evolutionary reactors could be ready for deployment by 2000, and the mid-sized could be ready for initial plant construction soon after 2000. The Committee's evaluations and overall assessment are summarized in Figure 5-1.

Reactor Designation	Available Design Information	Evaluation Criteria: Safety	Economy	Market Suitability	Fuel Cycle	Safeguards & Physical Sec.	Maturity of Development	Licensing	Overall Assessment[1]
ABWR[a]	○	○	○	○	◒	◒	○	○	○
APWR[a]	○	○	○	○	◒	◒	○	○	○
SYS 80+[a]	○	○	○	○	◒	◒	○	○	○
AP 600[b]	◒	○	◒	○	◒	◒	◒	○	○
SBWR[b]	◒	○	◒	○	◒	◒	◒	○	○
CANDU	○	○	●	◒	◒	◒	○	◒	◒
SIR	◒	○	◒[2]	●	◒	◒	●	●	●
MHTGR	◒	○	●	●	◒	◒	●	●	●
PIUS	●	○	◒[2]	●	◒	◒	●	●	●
PRISM[c] LMR	●	○	●[3]	●[3]	○	◒	●	●	◒

Legend Rating: ○ high ◒ moderate ● low

Notes: 1 Overall assessment was mostly driven by market suitability.
2 Lack of design maturity results in great uncertainty relative to vendor cost projections.
3 Long-term economy and market potential could be high, depending on uranium resource availability.
a ABWR - *advanced boiling water reactor;* APWR - *advanced pressurized water reactor;* and SYS 80+ - *system 80+. Large evolutionary LWRs.*
b AP 600 - *advanced pressurized 600;* and SBWR - *simplified boiling water reactor. Mid-sized LWRs with passive safety features.*
c PRISM - *power reactor, innovative small module.*

FIGURE 5-1 Assessment of advanced reactor technologies.

This table is an attempt to summarize the Committee's qualitative rankings of selected reactor types ***against each other***, without reference either to an absolute standard or to the performance of any other energy resource options. This evaluation was based on the Committee's professional judgment.

The Committee has concluded the following:

1. Safety and cost are the most important characteristics for future nuclear power plants.

2. LWRs of the large evolutionary and the mid-sized advanced designs offer the best potential for competitive costs (in that order).

3. Safety benefits among all reactor types appear to be about equal at this stage in the design process. Safety must be achieved by attention to all failure modes and levels of design by a multiplicity of safety barriers and features. Consequently, in the absence of detailed engineering design and because of the lack of construction and operating experience with the actual concepts, vendor claims of safety superiority among conceptual designs cannot be substantiated.

4. LWRs can be deployed to meet electricity production needs for the first quarter of the next century:

a. The evolutionary LWRs are further developed and, because of international projects, are most complete in design. They are likely to be the first plants certified by NRC. They are expected to be the first of the advanced reactors available for commercial use and could operate in the 2000 to 2005 time frame. Compared to current reactors, significant improvements in safety appear likely. Compared to recently completed high-cost reactors, significant improvements also appear possible in cost if institutional barriers are resolved. While little or no federal funding is deemed necessary to complete the process, such funding could accelerate the process.

b. Because of the large size and capital investment of evolutionary reactors, utilities that might order nuclear plants may be reluctant to do so. If nuclear power plants are to be available to a broader range of potential U.S. generators, the development of the mid-sized plants with passive safety features is important. These reactors are progressing in their designs, through DOE and industry funding, toward certification in the 1995 to 2000 time frame. The Committee believes such funding will be necessary to complete the process. While a prototype in the traditional sense will not be required, federal funding will likely be required for the first mid-sized LWR with passive safety features to be ordered.

c. Government incentives, in the form of shared funding or financial guarantees, would likely accelerate the next order for a light water plant. The Committee has not addressed what type of government assistance should be provided nor whether the first advanced light water plant should be a large evolutionary LWR or a mid-sized passive LWR.

5. The CANDU-3 reactor is relatively advanced in design but represents technology that has not been licensed in the United States. The Committee did not find compelling reasons for federal funding to the vendor to support the licensing.

6. SIR and PIUS, while offering potentially attractive safety features, are unlikely to be ready for commercial use until after 2010. This alone may limit their market potential. Funding priority for research on these reactor systems is considered by the Committee to be low.

7. MHTGRs also offer potential safety features and possible process heat applications that could be attractive in the market place. However, based on the extensive experience base with light water technology in the United States, the lack of success with commercial use of gas technology, the likely higher costs of this technology compared with the alternatives, and the substantial development costs that are still required before certification,[3] the Committee concluded that the MHTGR had a low market potential. The Committee considered the possibility that the MHTGR might be selected as the new tritium production reactor for defense purposes and noted the vendor association's estimated reduction in development costs for a commercial version of the MHTGR. However, the Committee concluded, for the reasons summarized above, that the commercial MHTGR should be given low priority for federal funding.

8. LMR technology also provides enhanced safety features, but its uniqueness lies in the potential for extending fuel resources through breeding. While the market potential is low in the near term (before the second quarter of the next century), it could be an important long-term technology, especially if it can be demonstrated to be economic. The Committee believes that the LMR should have the highest priority for long-term nuclear technology development.

9. The problems of proliferation and physical security posed by the various technologies are different and require continued attention. Special attention will need to be paid to the LMR.

Alternative Research and Development Programs

The Committee developed three alternative R&D programs, each of which contains three common research elements: (1) reactor research using federal facilities. The experimental breeder reactor-II, hot fuel examination facility/south, and fuel manufacturing facility are retained for the LMR; (2) university research programs; and (3) improved performance and life extension programs for existing U.S. nuclear power plants.

---

[3] The Gas Cooled Reactor Associates estimates that, if the MHTGR is selected as the new tritium production reactor, development costs for a commercial MHTGR could be reduced from about $1 billion to $0.3 - 0.6 billion.[DOE, 1990 in Chapter 3]

The Committee concluded that federal support for development of a commercial version of the MHTGR should be a low priority. However, the fundamental design strategy of the MHTGR is based upon the integrity of the fuel ($\leq 1600^{\circ}C$) under operation and accident conditions. There are other potentially significant uses for such fuel, in particular, space propulsion. Consequently, the Committee believes that DOE should consider maintaining a coated fuel particle research program within that part of DOE focused on space reactors.

Alternative 1 adds funding to assist development of the mid-sized LWRs with passive safety features. Alternative 2 adds a LMR development program and associated facilities--the transient reactor test facility, the zero power physics reactor, the Energy Technology Engineering Center, and either the hot fuel examination facility/north in Idaho or the Hanford hot fuel examination facility. This alternative would also include limited research to examine the feasibility of recycling actinides from LWR spent fuel, utilizing the LMR. Finally, Alternative 3 adds the fast flux test facility and increases LMR funding to accelerate reactor and integral fast reactor fuel cycle development and examination of actinide recycle of LWR spent fuel.

None of the three alternatives contain funding for development of the MHTGR, SIR, PIUS, or CANDU-3.

Significant analysis and research is required to assess both the technical and economic feasibility of recycling actinides from LWR spent fuel. The Committee notes that a study of separations technology and transmutation systems was initiated in 1991 by DOE through the National Research Council's Board on Radioactive Waste Management.

It is the Committee's judgment that Alternative 2 should be followed because it:

- provides adequate support for the most promising near-term reactor technologies;
- provides sufficient support for LMR development to maintain the technical capabilities of the LMR R&D community;
- would support deployment of LMRs to breed fuel by the second quarter of the next century should that be needed; and
- would maintain a research program in support of both existing and advanced reactors.

Individual Views

NEIL E. TODREAS
APRIL 30, 1991

The Committee's endorsement of policy Alternative #2 is an affirmation that focused sequential exploitation of light water reactor (LWR) development in the short term and liquid metal reactor (LMR) development for the long term should be pursued. Acceleration of liquid metal development by an additional ten years as envisioned in Alternative #3 was not deemed necessary at this time. Concurrent with rejection of this acceleration it was deemed possible to conduct liquid metal fuel irradiation technology development without the FFTF, albeit at some risk.

However, an assessment of this risk including options to minimize it were not explicitly detailed. The technical risk principally involves uncertainties in extrapolating short-length (13″) EBR-II fuel test specimen results to design conditions (53″). Current long specimens (36″) in FFTF will not be taken to design burnups and examined for some years to ensure that this risk of extrapolation is tolerable. Further, the continued availability of EBR-II, now twenty-eight years old, is not assured. The example of the French reactor RAPSODIE which was ultimately shutdown in 1983, following discovery the previous year of an unrepairable leak, testifies to the uncertainties in assuring the availability of aging test reactors.

It therefore seems prudent to the execution of the recommended LMR development strategy that the contemplated retirement of FFTF be done in stages including an initial approximately five-year mothball status period. This would allow the following evaluations to be completed to ensure that the retirement of FFTF was a cost effective decision.

1. Complete the destructive examinations of an adequate number of fuel specimens of long length (36″) taken to design burnups in both FFTF steady state and TREAT transient tests. This will allow assessment of the risk inherent in extrapolating the behavior of short-length (13″) fuel specimens upon which the irradiation tests program is almost exclusively to be based. This risk assessment will indicate the number of demonstration plants of increasing core size that will be needed to prudently reach prototype design conditions. Life cycle development costs are strongly affected by this judgment since the tradeoff is between a strategy of building multiple demonstration plants and retiring the FFTF versus a path of building perhaps a single intermediate-size demonstration plant and retention of the FFTF.

2. Perform a thorough evaluation of the maintenance and modernization costs to ensure as best as possible the continued availability of the EBR-II as an irradiation test bed for the LMR development program.

During this five-year period in which the determination of the U.S. to maintain an LMR program as a strategic national program is demonstrated, it might become possible to develop a complementary international LMR consensus. From such a consensus would almost certainly arise a cooperative fuel and safety program whose centerpiece would be the utilization of FFTF as the international test reactor. The resulting FFTF support costs for the U.S. could be significantly lower than those existing today. These reduced maintenance costs could then be considered in conjunction with the costs identified in (1) and (2) above to reach an informed judgment on the question of maintenance or retirement of the FFTF.

Adolf Birkhofer and Sol Burstein agree with this separate opinion.

HOWARD K. SHAPAR
MARCH 29, 1991

There are three issues on which I wish to state my separate views.

First, I disagree with the Committee's conclusion that there is need for the establishment of a new entity whose functions would parallel those of the National Transportation Safety Board. The stated bases for the Committee's conclusion are that the establishment of such an entity would enhance safety and public acceptance because, among other things, it would facilitate a determination of the NRC staff's role in contributing to accidents. To think that the establishment of such an accident-review body would have a significant impact on safety or public acceptance of nuclear power is naive in the extreme. The last thing this country needs is another bureaucracy to review nuclear safety issues. To the extent that staff involvement may be a contributing factor to the occurrence of a nuclear incident, I fail to see why NRC's Inspector General cannot be reasonably expected to bring that involvement to light. If it were to be shown (which I strongly doubt) that NRC's Inspector General could not perform this function, then any small review entity created should be based on the following:

- make the entity part of NRC and accord the Commission sufficient supervisory authority to assure that the entity's efforts complement rather than conflict with NRC's regulatory responsibilities;
- confine the entity's functions to investigative fact-finding, causal determination and reporting thereon;
- establish criteria for triggering safety investigations in order to assure a focus on matters of real safety significance; and

- assure that the technical resources of NRC will not be duplicated or diluted. The entity should have a small professional staff to provide investigative leadership, with the detailing of experts from NRC or elsewhere to fill particular investigative needs.

Second, the Committee's report should, in my view, contain a strong recommendation for the enactment of legislation, along the lines of the legislation put forward recently by the Administration in connection with its National Energy Strategy, which would minimize the possibility of a post-construction hearing in cases where a combined construction permit and operating license had been issued. One of the major obstacles to continued nuclear power development in the United States has been the failure of NRC's licensing process to provide utilities and investors with requisite levels of certainty that a completed plant will be allowed to begin commercial operation on schedule. While NRC has made substantial progress towards the goal of streamlining its regulatory process by issuing rules in April 1989 (10 CFR Part 52) to provide for early site permits, standardized design certifications, and combined construction permits and operating licenses, there remains the substantial possibility, if not likelihood, of another hearing (after the plant has been completed) which could lead to extensive delays before the plant could go into operation. What is needed is a nuclear power plant licensing process that will permit operation in five or six years as is routinely the case in France and Japan. Clearly, delays in nuclear plant construction and operation have a fundamental impact on electric generation costs. Recent experience shows that even when licensing hearings are focused on narrow, technical issues, it can take many months or years to resolve those issues. Given the cost of carrying a multi-billion dollar investment while these issues are being litigated, the uncertainties associated with a post-construction hearing are sufficiently daunting to deter utilities from ordering a nuclear plant. In short, legislation such as I would recommend would go a long way to increasing utility and investor confidence in future nuclear power plant orders.

Third, I do not believe that the Committee's report deals adequately with a disturbing trend -- the fact that an increasing (but, thus far, small) number of States are involving themselves in matters of nuclear safety. If this trend continues, another institutional barrier to the further development of nuclear power in the United States will have been erected. Two of the principal reasons why the Congress wisely decided that nuclear safety should be the exclusive responsibility of the Federal Government was the belief (1) that dual regulation by the States and the Federal Government would be counter-productive to safety and (2) that Federal preemption would better serve the development of nuclear energy in the United States. Those reasons continue to be valid today. The Committee's report should have concluded that entry by the States into the field of nuclear power plant safety is (in addition to being unlawful) ill-advised, counter-productive to safety, and contrary to the National interest.

Appendix A

COMMITTEE MEETINGS

First Meeting
May 31 - June 1, 1989
National Academy of Sciences
Washington, D.C.

Key Objectives: To review the task statement for the study; to hear presentations on DOE's current reactor development program and discuss relevant background material and perspectives; to identify the practical technological options and an appropriate proponent group for each; to develop evaluation criteria; and to determine future meeting dates and assignments.

Presentations:

Current DOE Reactor Development Program and Expectations for Study

Overview
Mary Ann Novak, Acting Assistant Secretary for Nuclear Energy

Advanced Light Water Reactor
David McGoff, Associate Deputy Assistant Secretary for Reactor Deployment

Liquid Metal Reactor
Jerry Griffith, Associate Deputy Assistant Secretary for Reactor Systems Development and Technology

High Temperature Gas Reactor
Jerry Griffith, Associate Deputy Assistant Secretary for Reactor Systems Development and Technology

Legislative Perspective and Expectations for Study
Ben Cooper, Senior Professional Staff
Senate Committee on Energy and Natural Resources

Overview of Congressional Office of Technology Assessment Study: Nuclear Power in an Age of Uncertainty
Alan Crane, Senior Associate
Congressional Office of Technology Assessment

Overview of Massachusetts Institute of Technology Study: National Strategies for Nuclear Power Development, prepared under National Science Foundation grant

Richard Lester, Professor, Department of Nuclear Engineering
Massachusetts Institute of Technology

Overview of U.S. Department of Energy's Energy Research Advisory Board Study: Review of the Proposed Strategic National Plan for Civilian Nuclear Reactor Development

John Landis, Sr. Vice President and Director
Stone & Webster

Dinner Speaker:

Llewellyn King, Publisher
The Energy Daily

Second Meeting
August 21-25, 1989
Beckman Center
Irvine, California

Key Objectives: To obtain information on the technological options from their proponents and to begin evaluating the technologies. This meeting was a Reactor Technologies Workshop at which the Committee members received presentations on various reactor technologies from U.S. and foreign speakers. Also during this meeting staff presented the results of interviews with various utility chief executive officers. Following the presentations, the Committee broke into focus groups for the evaluation.

Presentations:

EPRI Overview of Advanced LWRs and Comments on Other Advanced Technologies

John Taylor, Vice President for Nuclear Power
John DeVine, Program Manager, Advanced Light Water Reactor
Electric Power Research Institute

Comments on Types and Status of Reviews

Thomas Murley, Director, Office of Nuclear Reactor Regulation
U.S. Nuclear Regulatory Commission

Combustion Engineering Advanced LWR Technology

Shelby Brewer, President, Nuclear Power Businesses
Combustion Engineering

General Electric Advanced LWR Technology
Bertram Wolfe, Vice President and General Manager
General Electric Company

Westinghouse Advanced LWR Technology
Howard Bruschi, Director of AP600 Program
Westinghouse

PIUS LWR Technology
Cnut Sundqvist, Reactor Division
Kåre Hannerz, PIUS Project
ABB Atom

Heavy Water Reactor Technology
D. R. Shiflett, Vice President and General Manager
AECL Technologies

Gas Cooled Reactor Technology
Linden Blue, Vice Chairman
General Atomics

John Jones, Director of Engineering Technology Division
Frank Homan, Director, Reactor Programs
Oak Ridge National Laboratory

Dan Mears, General Manager
Gas-Cooled Reactor Associates

Liquid Metal Reactor Technology
Bertram Wolfe Vice President and General Manager
Robert Berglund, Manager, Advance Nuclear Technology
General Electric

Charles Till, Associate Laboratory Director, Engineering Research
Yoon Chang, General Manager, IFR Program
Argonne National Laboratory

Discussion of Other Promising Technologies
Steven Hall
AEA Technology, United Kingdom

Third Meeting
October 19-20, 1989
National Academy of Sciences
Washington, D.C.

Key Objectives: To achieve consensus on the evaluation of technologies; to review progress on the analysis of institutional issues and decide on next steps; to set in motion plans for Tasks 4 and 5 (Development Approaches and Facilities); to meet with Admiral Watkins; and to complete the bias discussions.

Presentations:

U.S. Department of Energy Contracts and Facilities
Jerry Griffith, Acting Assistant Secretary for Nuclear Energy
Ray Hunter, Director, Office of Facilities, Fuel Cycle and Test Programs
U.S. Department of Energy

Dinner Speaker:

Admiral James Watkins, Secretary of Energy
U.S. Department of Energy

Fourth Meeting
December 7-8, 1989
Beckman Center
Irvine, California

Key Objectives: To receive the focus groups' reports on Tasks 4 and 5 (Development Approaches and Facilities) and establish the Committee's position on these tasks; to focus the dialogue on the institutional issues and establish the Committee's position on them; to clarify the issue on size (600 MWe versus 1000 MWe); to review a conceptual outline of the final report; to complete the evaluation of technologies (Task 3) by determining the Committee's views regarding development of the HTGR; to make assignments for the next meeting in January (Task 7: Policy Alternatives and Recommendations); and to receive a background briefing on global CO_2 emissions.

Presentation:

Carbon Dioxide and Global Warming
George Hidy, Vice President
Electric Power Research Institute

Fifth Meeting
January 29-30, 1990
Beckman Center
Irvine, California

Key Objectives: To reach preliminary consensus on the answers to the main questions that must be addressed in the Committee's final report; to isolate those issues requiring more analysis; and to adopt a schedule for completion of the study.

U.S. Department of Energy Presentations on Actinide Recycle and Facility Requirements

Speakers:

Jerry Griffith, Acting Assistant Secretary for Nuclear Energy
Sol Rosen, Director of Advanced Reactor Programs
Clifford Weber, Senior Technical Assistant
Department of Energy

Robert Berglund, Manager, Advance Nuclear Technology
General Electric

William Burch, Director, Fuel Recycle Division
Oak Ridge National Laboratory

Yoon Chang, General Manager, IFR Program
Argonne National Laboratory

James Holmes, Manager, Development Department
Westinghouse Hanford Co.

Sixth Meeting
March 14-16, 1990
National Academy of Sciences
Washington, D.C.

Key Objectives: To review the first draft of the Committee's final report and arrive at consensus about what the report should say so it can be redrafted for Committee approval.

Presentations:

Proliferation Panel:

Harold A. Feiveson, School of Engineering/Applied Science, Center for Energy and Environmental Studies
Robert Williams, Center for Energy and Environmental Studies
Princeton University

Marvin Miller, Department of Nuclear Engineering
Massachusetts Institute of Technology

John Jones, Director of Engineering Technology Division
Oak Ridge National Laboratory

American Nuclear Energy Council:

Edward Davis, President

Dinner Speaker:

Kenneth Carr, Chairman
U.S. Nuclear Regulatory Commission

Seventh Meeting
August 8-10, 1990
National Academy of Sciences
Washington, D.C.

Key Objective: To reach closure on the Committee's final report so it can be sent to peer review.

Eighth Meeting
March 4-6, 1991
National Academy of Sciences
Washington, D.C.

Key Objective: To complete the final report so it can enter the Academy's peer review process.

Appendix B

CRITERIA DEVELOPED FOR THE COMPARATIVE ANALYSIS OF ADVANCED REACTOR TECHNOLOGIES

In accordance with the Statement of Task (see Preface, Task 2) the Committee developed criteria to evaluate the technological options. These criteria reflected the characteristics that the Committee deemed most important for future U.S. nuclear power plants. These criteria also were furnished to the reactor vendors before their presentations to the Committee. The Committee then assessed the relative merits of the reactor technologies under each broad heading (Criteria A through H) listed below.

A. SAFETY IN OPERATION

1. Safety Goal - Please identify the safety goals of the technology under discussion, why they were chosen, and the measures to achieve them. The discussion should include estimated probabilities of major core damage and release of radiation outside the plant, and the methodology and assumptions for the analysis. An explicit treatment of uncertainties should be included.

2. Safety Features - Explain various safety features--categorized into passive or active systems--that contribute to the total nuclear plant safety. Please identify the major engineering or scientific questions about the effectiveness of these features as well as the experiments or additional analyses needed to validate them.

3. Safety Issues - Please discuss how the proposed technology addresses various safety issues and dominant accident sequences. Additional safety concerns or accident sequences engendered by the technology should be treated. The one or two dominant safety issues should be identified.

4. Safety Indicators - Please discuss other potential safety indicators, such as estimated frequency of unplanned scrams.

5. Ease of Maintenance - Please discuss features of the technology that support or detract from ease of essential maintenance. Also, address the requirements and schedule for preventive maintenance.

6. Worker Safety - Safety of the workers during maintenance or otherwise should be addressed, along with expected occupational doses.

B. ECONOMY OF CONSTRUCTION AND OPERATION

1. Construction Costs - Please provide estimates of construction costs in constant 1989 dollars (indicating whether or not the costs of borrowing money are included in the estimate, and if so, what such costs are), and provide total nuclear plant costs as well as cost per installed electrical kilowatt. Please discuss how the estimates were arrived at, covering the principal assumptions, numerical estimates and so forth.

2. Time to Construct - Please provide estimates of the time likely to be required to construct a new nuclear plant and reach commercial operation. Discuss the estimate, including the assumptions made concerning the new Nuclear Regulatory Commission licensing process.

3. Uncertainties - Please estimate and discuss the uncertainties in both costs and time to reach commercial operation.

4. Operating Costs - Please provide estimates of operating costs over the plant's life expressed in constant 1989 dollars. Fuel cycle costs, operation and maintenance costs, and capital costs should be identified separately. Outline any assumptions you have made regarding new regulatory requirements. Expected lifetime of the plant should be discussed.

5. Availability - Please provide and discuss estimates of expected availability, and the frequency and duration of planned outages and refueling outages.

C. SUITABILITY FOR FUTURE MARKETS:

Your choices of technology, reactor characteristics and your organizational commitment to particular technologies were driven in part by perceptions of the character of future markets for new electrical generation equipment. Please discuss your organization's perceptions of the characteristics of this market, and discuss how your technology and reactor concepts fit these perceptions. For example:

1. Problems in the "First Nuclear Era" - Outline what you think went wrong with the current generation of power reactors and why the technology under discussion will not encounter the same obstacles.

2. Competitive Non-Nuclear Technologies - What technologies will offer the strongest competition to nuclear power in the future? How well will the nuclear technology being discussed fare in the most important dimensions of this competition?

3. Operational - What electrical demand growth rates are likely? What are the implications of current trends to deregulation of electricity generation

and to growth of independent power producers? How does the nuclear technology fit in the new framework you foresee? What revised mix of generating technologies seems most likely? What will be the role of nuclear? Can it and your technology meet revised demands such as load following?

4. Life Cycle Relationships - Address contemplated relationships--including lines of communication, authority, and responsibility as well as warranties--between organizations such as architect-engineers, vendors, nuclear steam system suppliers, owners, and operators from the design stage through decommissioning of a nuclear plant embodying the technology.

5. Financial - What attitudes do you foresee in the financial community to nuclear power and to your technology in particular? Indicate why the financial community would be inclined to lend money for this type of technology. For example, show how this technology can be expected to resolve some of the concerns about nuclear power, and how the technology would change the risk-reward ratio to make it more attractive than it is today for nuclear power.

6. Other Institutional Factors - Please address the technology in light of the competence of utility management (including the "utility management culture," a term referring to the quality and motivations of both management and staff that operate nuclear plants), the "regulatory compact" (State Public Utilities Commissions, Federal Energy Regulatory Commission, and prudency), emerging trends toward deregulation of electricity generation and acquisition of new power supply through competitive mechanisms, standardization (multiple architect/engineers, vendors, designers, and constructors), U.S. indemnity measures (such as Price-Anderson), and U.S. tax and incentive policies (Public Utility Regulatory Policies Act, Public Utility Holding Company Act).

7. Siting - Discuss siting needs, possible restrictions, and any other related issues.

D. FUEL CYCLE AND ENVIRONMENTAL CONSIDERATIONS: Please address the long-term environmental implications of the fuel cycle employed by the technology being presented. Specifically speak to the problems of spent fuel storage and long term waste disposal. Also comment on enrichment requirements. Please discuss "day-to-day" plant-related environmental considerations.

E. RESISTANCE TO DIVERSION AND SABOTAGE

1. Diversion - Please show how the technology addresses concerns about the possible forceful theft of militarily significant nuclear materials or their clandestine diversion. Address such events both at the nuclear plant itself and in the process of transporting nuclear materials to or from the plant.

2. Sabotage - Describe the features that could help prevent sabotage by outsiders or knowledgeable insiders.

F. TECHNOLOGY RISK AND DEVELOPMENT SCHEDULE

1. Critical Developments - Please identify the critical technical developments needed for this technology, and highlight the "go - no go" issues; indicate the essential elements of the R&D program to address these important matters. Identify the tests necessary to demonstrate critical technologies before commercialization.

2. Status of Tests and Demonstrations - Indicate what major systems or components have already been tested or demonstrated to a level of assurance believed adequate for use in future plants, and what remains to be tested or demonstrated.

3. Demonstration - Please explain the difficulty or ease of demonstrating whether the technology can live up to claims such as safety, economy, environmental acceptability, and licenseability. For example, would a demonstration plant have to be built to attain (or convince other important decision makers of the attainment of) the level of assurance believed necessary before deploying many such plants? How long would it have to be operated, and so forth?

4. Technology Readiness - Please estimate the date of availability of this technology for demonstration, and the date of availability for commercialization based on Nuclear Regulatory Commission certification as a standardized plant.

5. Research and Development Facility Requirements - Discuss the research and development (R&D) facilities needed. If U.S. government assistance is contemplated, identify what facilities (e.g., at the National Laboratories) would be needed to support R&D of the technology and the scope and estimated cost of such support.

6. Availability of Materials and Components - Indicate whether there are any critical materials, components, manufacturing processes, or other items

that are not presently available but would be necessary for successful commercialization. Discuss the resolution of any availability problems.

7. Critical Path Schedule - Please show and discuss a "master" or "top-level" schedule for completion of R&D, demonstration (if required) and commercialization, indicating the critical path items.

8. Development Cost - Provide a time phased estimate of the overall research, development, and demonstration costs (in constant 1989 dollars) needed to move from today's level of maturity to readiness for commercial application. Indicate the contemplated or potential source(s) for the money.

G. AMENABILITY TO EFFICIENT AND PREDICTABLE LICENSING

1. Efficiency - Discuss the ease or difficulty of meeting current or anticipated U.S. safety, environmental, and security and safeguards regulatory requirements (e.g., indicate whether the requirement for a containment would have to be changed).

2. Predictability - Identify any features of the technology that make it more likely to be licensed in a predictable way (e.g., simplicity and standardization). Identify any plausible regulatory requirements that could prevent the licensing of this technology. Address ways of resolving the "as licensed" versus "as built" issue.

H. NET ASSESSMENT: In closing please discuss, why, in your judgment the technology will be good enough to:

- cause the CEOs of utilities and of independent power production companies to buy it;
- give the financial community the reasons and the confidence to finance it; and
- cause the public to accept it.

List of Acronyms

ABWR	Advanced boiling water reactor
AECL	Atomic Energy of Canada Limited
APWR	Advanced pressurized water reactor
ALMR	Advanced liquid metal reactor
ATR	Advanced test reactor
BWR	Boiling water reactor
CANDU	Canadian deuterium uranium
CE	Combustion Engineering
DOE	U.S. Department of Energy
EBR-I	Experimental breeder reactor - I
EBR-II	Experimental breeder reactor - II
ECCS	Emergency core cooling system
EIA	Energy Information Administration
EPA	Environmental Protection Agency
EPRI	Electric Power Research Institute
ETEC	Energy Technology Engineering Center
ETR	Engineering test reactor
FFTF	Fast flux test facility
FMF	Fuel manufacturing facility
GE	General Electric
GA	General Atomics
HFEF/N	Hot fuel examination facility/north
HFEF/S	Hot fuel examination facility/south
HWR	Heavy water reactor
IFR	Integral fast reactor
INPO	Institute of Nuclear Power Operations
kWe	Kilowatt electric
kWh	Kilowatt hour
LWR	Light water reactor
LMR	Liquid metal reactor
MHTGR	Modular high-temperature gas-cooled reactor
MWe	Megawatts electric
MRS	Monitored retrievable storage
MTR	Materials test reactor
NPOC	Nuclear Power Oversight Committee
NRC	U.S. Nuclear Regulatory Commission
NTSB	National Transportation Safety Board
O&M	Operations and maintenance
OECD	Organization for Economic Cooperation and Development
ORNL	Oak Ridge National Laboratory
PIUS	Process inherent ultimate safety
PRA	Probabilistic risk analysis
PRISM	Power reactor, innovative small module

PUC	Public Utilities Commission
PURPA	Public Utility Regulatory Policies Act
R&D	Research and development
PWR	Pressurized water reactor
SALP	Systematic assessment of licensee performance
SBWR	Simplified boiling water reactor
SIR	Safe integral reactor
THTR	Thorium high-temperature reactor
TMI	Three Mile Island
TREAT	Transient reactor test
TVA	Tennessee Valley Authority
WANO	World Association of Nuclear Operators
ZPPR	Zero power physics reactor